RISING STARS
Mathematics

Year
6

Concept developed by
Caroline Clissold and Cherri Moseley

Year 6 Author Team
Caroline Clissold, Heather Davies,
Linda Glithro, Steph King

Pupil Textbook

The Publishers would like to thank the following for permission to reproduce copyright material.

Photo credits
Pages 10-11: skittles; pizzas – Shutterstock; page 23: remote control – Shutterstock; pages 25-6: calculator (right) – Shutterstock; pages 36-7: daffodils; paving; petrol pump; cakes; jug and glasses - Shutterstock; pages 50-1: lift; shapes; map; food packaging – Shutterstock; page 61 – football; fullerene – Shutterstock; pages 62-3: dials; stopwatch – Shutterstock; pages 72-3: money; clocks – Shutterstock; pages 86-7: egg boxes - Shutterstock; road sign; football pitch – iStock; pages 100-1: clothes – iStock; electrician; insect; parking meter – Shutterstock; pages 114-15: clock; Russian dolls – Shutterstock; square – bebsy/Shutterstock; parcels – iStock; page 125: crop circle – Hansueli Krapf/Wikipedia Commons; pages 126-7: scales and weights; apples – Shutterstock; swimming pool – Paval I Photo and Video/Shutterstock; page 135: child conjuror – Shutterstock; pages 136-7: planets; javelin track – Shutterstock; hats; eggs– iStock; pages 148-9: trees; buildings – Shutterstock; pages 160-1: measuring cups; children; market; beach – Shutterstock; calendar – iStock; pages 172-3: columns; footballers – Shutterstock; figures; reflection – iStock; docks – Christian Mueller/Shutterstock; page 183: hot air balloons – Shutterstock.

Acknowledgements
The reasoning skills on page 8 are based on John Mason's work on mathematical powers. See Mason, J. and Johnston-Wilder, S. (Eds.) (2004). Learners powers. *Fundamental constructs in Mathematics Education*. London: Routledge Falmer. 115-142.

Every effort has been made to trace all copyright holders, but if any have been inadvertently overlooked, the Publishers will be pleased to make the necessary arrangements at the first opportunity.
Although every effort has been made to ensure that website addresses are correct at time of going to press, Rising Stars cannot be held responsible for the content of any website mentioned in this book. It is sometimes possible to find a relocated web page by typing in the address of the home page for a website in the URL window of your browser.

Hachette UK's policy is to use papers that are natural, renewable and recyclable products and made from wood grown in sustainable forests. The logging and manufacturing processes are expected to conform to the environmental regulations of the country of origin.

ISBN: 978 1 78339 527 9
Text, design and layout © Rising Stars UK Ltd 2016
First published in 2016 by
Rising Stars UK Ltd, part of Hodder Education,
An Hachette UK Company
Carmelite House
50 Victoria Embankment
London EC4Y 0DZ
www.risingstars-uk.com
Authors: Caroline Clissold, Heather Davies, Linda Glithro, Steph King

Programme consultants: Caroline Clissold, Cherri Moseley, Paul Broadbent
Publishers: Fiona Lazenby and Alexandra Riley
Editorial: Kate Baxter, Jane Carr, Jan Fisher, Lucy Hyde, Lynette James, Jackie Mace, Jane Morgan, Christine Vaughan
Project manager: Sue Walton
Series and character design: Steve Evans
Illustrations by Steve Evans

Cover design: Steve Evans and Words & Pictures
Printed by Liberduplex, Barcelona
A catalogue record for this title is available from the British Library.

Contents

Introduction

Hello, I'm Eva. Welcome to *Rising Stars Mathematics!*

Look at the pictures at the beginning of the unit. Think about the mathematics you can see in the world around you.

Talk about the questions with your friends. Do you agree on the answers?

Read what Eva and Ali say. Can you spot if they have made a mistake?

Read the text and look at the diagrams to learn new maths skills. Your teacher will explain them.

Do these activities to practise what you have learnt. Write the answers in your exercise book.

These questions will help you explore and investigate maths. You will need to think about them carefully.

Use these items to help you. Make sure you have everything you need.

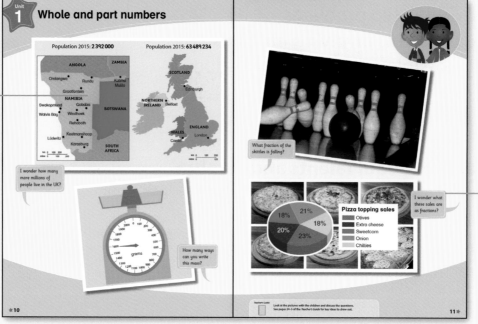

And I'm Ali. We'll help you as you learn with this book!

Play the game at the end of the unit to practise what you have learnt.

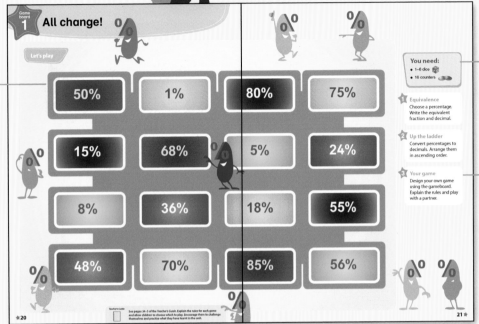

Game board 1 — All change!

Let's play

50%	1%	80%	75%
15%	68%	5%	24%
8%	36%	18%	55%
48%	70%	85%	56%

You need:
- 1–6 dice
- 16 counters

1 Equivalence
Choose a percentage. Write the equivalent fraction and decimal.

2 Up the ladder
Convert percentages to decimals. Arrange them in ascending order.

3 Your game
Design your own game using the gameboard. Explain the rules and play with a partner.

See pages 14–5 of the Teacher's Guide. Explain the rules for each game and allow children to choose which to play. Encourage them to challenge themselves and practise what they have learnt in the unit.

Make sure you have everything you need.

Follow the instructions to use the gameboard in different ways.

20   21

Try these activities to check what you have learnt in the unit. Have you understood all the new maths concepts?

Review 1 — And finally …

Let's review

1 These 3 numbers are in descending order:
72 456.91, 34 822.6, 96 257

Explain where Eva has gone wrong and why.

2

| 8 | 4 | 5 | 1 | 9 | . |

Use the digits to make up a 3-digit number with 2 decimal places. Underneath your number, write what it would be after it is:

a   Multiplied by 10

b   Multiplied by 100

c   Divided by 10

d   Divided by 100

Round all your new numbers to the nearest whole number.

Repeat with another 3-digit number with 2 decimal places.

See pages 26–7 of the Teacher's Guide for guidance on running each task. Observe children to identify those who have mastered concepts and those who require further consolidation.

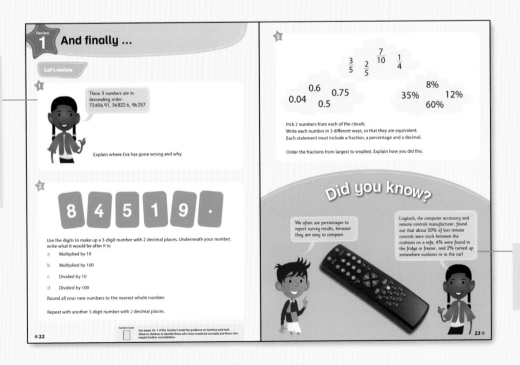

3

$\frac{3}{5}$ $\frac{2}{5}$ $\frac{7}{10}$ $\frac{1}{4}$

0.6   0.75   8%   12%
0.04   0.5   35%   60%

Pick 2 numbers from each of the clouds.
Write each number in 3 different ways, so that they are equivalent.
Each statement must include a fraction, a percentage and a decimal.

Order the fractions from largest to smallest. Explain how you did this.

Did you know?

We often use percentages to report survey results, because they are easy to compare.

Logitech, the computer accessory and remote controls manufacturer, found out that about 50% of lost remote controls were stuck between the cushions on a sofa, 4% were found in the fridge or freezer, and 2% turned up somewhere outdoors or in the car!

Find out more about maths by reading these fun facts!

22   23

Problem solving and reasoning

Try these ideas to develop your reasoning skills. Doing this will help you improve your mathematical thinking.

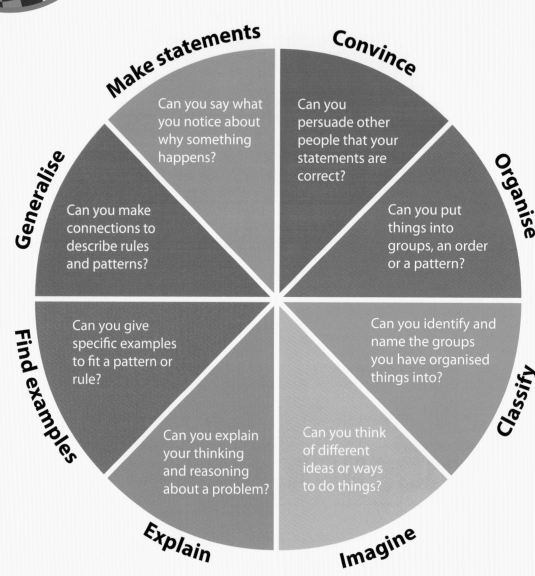

Make statements
Can you say what you notice about why something happens?

Convince
Can you persuade other people that your statements are correct?

Generalise
Can you make connections to describe rules and patterns?

Organise
Can you put things into groups, an order or a pattern?

Find examples
Can you give specific examples to fit a pattern or rule?

Classify
Can you identify and name the groups you have organised things into?

Explain
Can you explain your thinking and reasoning about a problem?

Imagine
Can you think of different ideas or ways to do things?

Follow these steps to help you solve problems!

1 Read the problem carefully.

2 What do you need to find out?

3 What data or information is given in the problem?

4 What data or information do you need to use?

5 Make a plan for what to do.

6 Follow your plan to find the answer.

7 Check your answer. Is it correct? Put your answer into the problem to see if it works with the information given.

8 Evaluate your method. How could you improve it next time?

Whole and part numbers

Population 2015: **2 392 000**

Population 2015: **63 489 234**

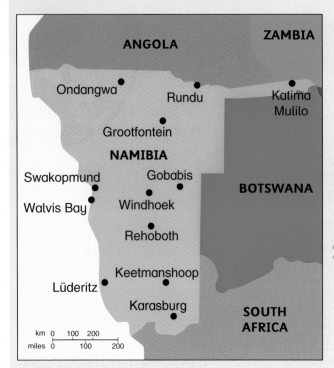

I wonder how many more millions of people live in the UK?

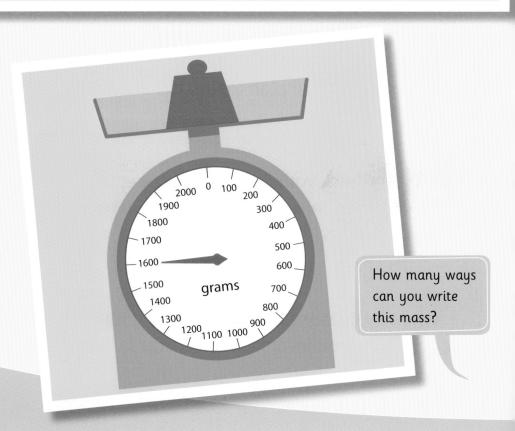

How many ways can you write this mass?

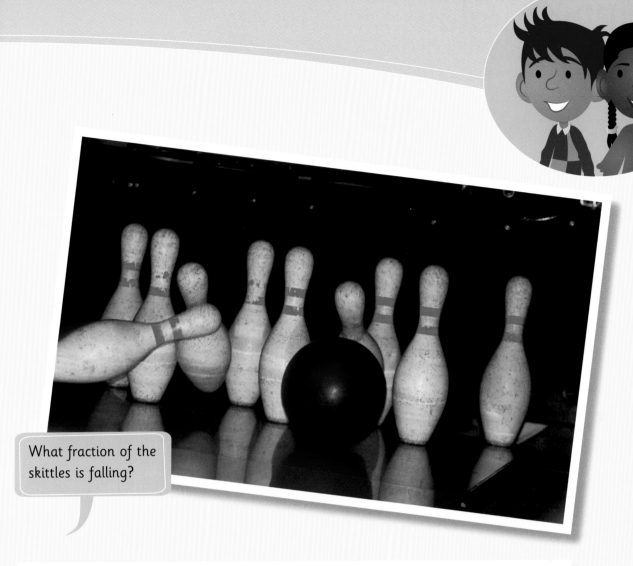

What fraction of the skittles is falling?

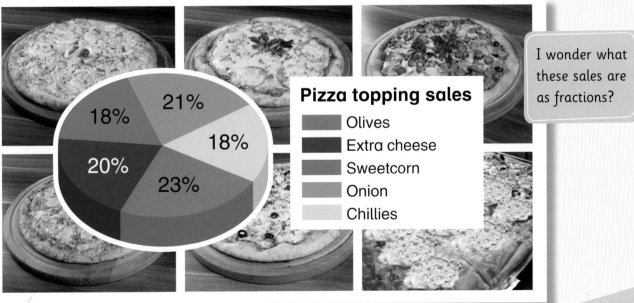

Pizza topping sales

	Olives
	Extra cheese
	Sweetcorn
	Onion
	Chillies

21% 18% 18% 20% 23%

I wonder what these sales are as fractions?

Teacher's Guide
Look at the pictures with the children and discuss the questions.
See pages 24–5 of the *Teacher's Guide* for key ideas to draw out.

11 ★

Place value

You need:
- digit cards
- place-value grid
- money (coins and notes)

When you multiply by 10 and 100, you just add zeros.

It might look like that when you multiply whole numbers but it doesn't work for all numbers. If you just added zeros, 1.6 × 100 would be 1.600, and that's the same number.

Place value

1 000 000	100 000	10 000	1000	100	10	1	.	$\frac{1}{10}$	$\frac{1}{100}$
			3	8	2	7	.	5	
				5	8	2	.	7	5
		5	8	2	7	5	.		

3 0 0 0 .0 ▷   2 0 .0 ▷   .5 ▷

8 0 0 .0 ▷   7 .0 ▷

In the place-value grid the digit 3 is in the thousands position. To get its true value you multiply it by 1000.

Look at the remaining digits in the first row of the place-value grid. Can you identify their position and true value?

Now you have the values of all the parts of the number, you add them together to get the whole number:

$3000 + 800 + 20 + 7 + \frac{5}{10} = 3827.5$

Multiplying and dividing by 10 and 100

When you multiply by 10, you make each digit 10 times bigger.

When you divide by 10, you make each digit 10 times smaller.

When you multiply by 100, you make each digit 100 times bigger.

When you divide by 100, you make each digit 100 times smaller.

What has happened to the numbers in the second and third rows of the place-value grid?

Teacher's Guide

Before working through the *Textbook*, study page 26 of the *Teacher's Guide* to see how the concepts should be introduced. Read and discuss the page with the children. Provide concrete resources to support exploration.

★12

1

Work out.

Write the position and value of the 6 in these numbers:

a 12 469	c 452 456
b 262 985	d 3 245 652

Write the position and value of the 5 in these numbers:

e 172 156	g 127 501
f 512 386	h 2 815 210

2

Calculate.

Multiply these by 10.

a 12	c 10.2
b 7.5	d 15.8

Divide these by 10.

e 14	g 8
f 21	h 5

3

Apply.

Make these amounts using the fewest coins.

a 3p	c 21p
b 15p	d 49p

Multiply the amounts by 10 and 100.
Show each new amount using the fewest coins.

Make these amounts using the fewest notes and coins.

e £2	g £11
f £4	h £13

Divide the amounts by 10 and 100.
Show each new amount using the fewest coins.

4

Think.

On a calculator Luke keyed in a number. He multiplied or divided it by 10 or 100.
The result was 23.48. What number could he have started with and what might he have done to the number?

Find another 3 possibilities.

Teacher's Guide

See page 27 of the *Teacher's Guide* for ideas of how to guide practice.
Work through each step together as a class to develop children's conceptual understanding.

13 ★

Comparing, ordering and rounding numbers

Let's learn

You need:
- digit cards
- place-value grid
- money (coins and notes)

Ordering numbers is easy. You just look at the first digit and the highest one means you have the highest number.

That sometimes works, but not if you have numbers with the same first digit.

Ordering and comparing numbers

Ascending order means ordering from the lowest to the highest number:

234 785, 234 885, 234 965.

Descending order means ordering from the highest to the lowest number:

428 734, 428 334, 428 128.

The numbers in the place-value grid all begin with the same 4 digits. Which digits do you need to look at to order the numbers?

1 000 000	100 000	10 000	1000	100	10	1	.	$\frac{1}{10}$	$\frac{1}{100}$
1	5	6	0	7	2	3	.	5	
1	5	6	0	5	2	3	.	7	5
1	5	6	0	9	2	3	.	1	2

Rounding numbers

Gattegno chart:

0.01	0.02	0.03	0.04	0.05	0.06	0.07	0.08	0.09
0.1	0.2	0.3	0.4	0.5	0.6	0.7	0.8	0.9
1	2	3	4	5	6	7	8	9
10	20	30	40	50	60	70	80	90
100	200	300	400	500	600	700	800	900

To round a number:

- Decide which place value you are rounding to, e.g. the nearest 10, the nearest 100, the nearest 1000.

- If the next digit in the number is 5 or more, round up.
- If it is less than 5, leave it the same.

Teacher's Guide

Before working through the *Textbook*, study page 28 of the *Teacher's Guide* to see how the concepts should be introduced. Read and discuss the page with the children. Provide concrete resources to support exploration.

1 Order these.

In ascending order:

a   4572, 4578, 4571, 4575, 4579

b   394 125, 394 178, 394 234, 394 215, 394 193

In descending order:

c   2832, 2945, 2846, 2936, 2935

d   1 354 287, 1 358 123, 1 359 346, 1 358 154, 1 354 217

2 Round.

Use the Gattegno chart to make up a 3-digit number with 2 decimal places.
Round it to the nearest:

a   tenth   b   whole number   c   ten   d   hundred

3 Investigate.

Make these amounts of money using the fewest number of notes and coins.

a   £4.97   b   £8.13   c   £10.56   d   £17.62   e   £22.25   f   £26.12

Now round them to the nearest pound. Solve these problems. Round up 1 amount to a multiple of 10 by taking what you need from the other amount.

g   Charlie bought a DVD player for £115 and a flat screen TV for £249. How much did he spend in total?

h   Nafisat spent £12.99 on a book and £54.50 on a DVD box set. How much did she spend?

4 Think.

Use the Gattegno chart to make up two 3-digit numbers with 2 decimal places.
Compare them.
Write down 4 ways in which they are the same.

Write down 4 ways in which they are different.
Repeat with another pair of 3-digit numbers with 2 decimal places.

Teacher's Guide

See page 29 of the *Teacher's Guide* for ideas of how to guide practice.
Work through each step together as a class to develop children's conceptual understanding.

15 ★

Comparing, ordering and simplifying fractions

Let's learn

You need:
- strips of paper

$\frac{1}{10}$ is obviously bigger than $\frac{1}{5}$ because 10 is bigger than 5.

$\frac{1}{5}$ is bigger. Imagine 2 bars of chocolate the same size. If you break 1 into tenths, that will be 10 pieces. If you break the other into fifths that will be 5 pieces. The fifths are bigger than the tenths!

Comparing and ordering fractions

Sometimes you need to compare and order fractions according to their size.

For unit fractions, the smaller the denominator, the larger the fraction.

Look at the fraction strips. Can you see that $\frac{1}{2}$ is bigger than $\frac{1}{4}$ and $\frac{1}{8}$?

1									
$\frac{1}{2}$					$\frac{1}{2}$				
$\frac{1}{5}$		$\frac{1}{5}$		$\frac{1}{5}$		$\frac{1}{5}$		$\frac{1}{5}$	
$\frac{1}{10}$	$\frac{1}{10}$	$\frac{1}{10}$	$\frac{1}{10}$	$\frac{1}{10}$	$\frac{1}{10}$	$\frac{1}{10}$	$\frac{1}{10}$	$\frac{1}{10}$	$\frac{1}{10}$

1															
$\frac{1}{2}$								$\frac{1}{2}$							
$\frac{1}{4}$				$\frac{1}{4}$				$\frac{1}{4}$				$\frac{1}{4}$			
$\frac{1}{8}$		$\frac{1}{8}$		$\frac{1}{8}$		$\frac{1}{8}$		$\frac{1}{8}$		$\frac{1}{8}$		$\frac{1}{8}$		$\frac{1}{8}$	
$\frac{1}{16}$	$\frac{1}{16}$	$\frac{1}{16}$	$\frac{1}{16}$	$\frac{1}{16}$	$\frac{1}{16}$	$\frac{1}{16}$	$\frac{1}{16}$	$\frac{1}{16}$	$\frac{1}{16}$	$\frac{1}{16}$	$\frac{1}{16}$	$\frac{1}{16}$	$\frac{1}{16}$	$\frac{1}{16}$	$\frac{1}{16}$

Simplifying fractions

You can simplify fractions to make calculations easier.

Find a common factor of the numerator and denominator and divide it into them.

Look at $\frac{75}{100}$. A common factor of 75 and 100 is 25. So $\frac{75}{100}$ is equivalent to $\frac{3}{4}$.

It is much easier to find $\frac{3}{4}$ of a number than $\frac{75}{100}$!

Teacher's Guide

Before working through the *Textbook*, study page 30 of the *Teacher's Guide* to see how the concepts should be introduced. Read and discuss the page with the children. Provide concrete resources to support exploration.

1

Answer these.

Order these fractions from smallest to largest.

a $\frac{1}{4}, \frac{1}{8}, \frac{1}{10}, \frac{1}{2}, \frac{1}{5}$

b $\frac{1}{9}, \frac{1}{6}, \frac{1}{2}, \frac{1}{3}, \frac{1}{12}$

c $\frac{1}{3}, \frac{1}{5}, \frac{1}{8}, \frac{1}{10}, \frac{1}{4}$

d $\frac{1}{7}, \frac{1}{16}, \frac{1}{5}, \frac{1}{2}, \frac{1}{12}$

Write 2 number statements for each pair of fractions using the symbols > and <.

e $\frac{3}{5}$ and $\frac{5}{8}$

f $\frac{2}{3}$ and $\frac{7}{9}$

g $\frac{3}{4}$ and $\frac{7}{10}$

h $\frac{11}{12}$ and $\frac{3}{4}$

2

Simplify.

a $\frac{20}{25}$　　b $\frac{30}{36}$　　c $\frac{55}{99}$　　d $\frac{48}{72}$　　e $\frac{72}{108}$　　f $\frac{60}{80}$

3

Apply.

Take 8 strips of paper 10 cm in length. Fold or measure them and shade these fractions. Write the length of each fraction in the shaded part.

a $\frac{1}{2}$　　　　　e $\frac{1}{10}$

b $\frac{1}{4}$　　　　　f $\frac{3}{10}$

c $\frac{3}{4}$　　　　　g $\frac{4}{5}$

d $\frac{1}{5}$　　　　　h $\frac{7}{10}$

What would the values of the fractions be if the whole strip represented 1 kg?

What if the whole strip represented £5?

4

Think.

Dan simplified some fractions. He ended up with $\frac{2}{5}$ and $\frac{7}{8}$.

Both denominators were multiples of 40.

Write down 5 possibilities of what his original fractions could be.

First, discuss which methods you could use with a partner.

Teacher's Guide

See page 31 of the *Teacher's Guide* for ideas of how to guide practice. Work through each step together as a class to develop children's conceptual understanding.

17

1d Equivalences

You need:
- paper strip
- tangram
- 10 × 10 square grids
- containers
- measuring jugs

Let's learn

We were drawing tangrams in class today. Someone said that the parallelogram and the square are the same fraction of the tangram. They can't be because they don't look the same.

It doesn't matter what they look like. Fractions of shapes are all about area, not shape.

Comparing fractions within shapes

To compare fractions within shapes you need to look at the areas the shapes take up.

Look at the 2 small triangles in the tangram. They would fit inside the square and the parallelogram perfectly. So the square and the parallelogram both have the same area – they are equivalent.

Equivalences between fractions, decimals and percentages

Fractions, decimals and percentages are all about part–whole relationships.

Percentages are parts of 100. So 25% is 25 out of 100, which we can write as $\frac{25}{100}$. We can also say that 25% is 25 hundredths, or 0.25.

Look at the shaded parts of the percentage grid. What would these be as percentages, fractions and decimals?

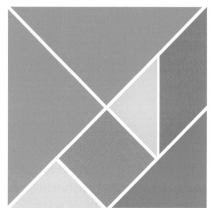

1%	1%	1%	1%	1%	1%	1%	1%	1%	1%
1%	1%	1%	1%	1%	1%	1%	1%	1%	1%
1%	1%	1%	1%	1%	1%	1%	1%	1%	1%
1%	1%	1%	1%	1%	1%	1%	1%	1%	1%
1%	1%	1%	1%	1%	1%	1%	1%	1%	1%
1%	1%	1%	1%	1%	1%	1%	1%	1%	1%
1%	1%	1%	1%	1%	1%	1%	1%	1%	1%
1%	1%	1%	1%	1%	1%	1%	1%	1%	1%
1%	1%	1%	1%	1%	1%	1%	1%	1%	1%
1%	1%	1%	1%	1%	1%	1%	1%	1%	1%

Teacher's Guide

Before working through the *Textbook*, study page 32 of the *Teacher's Guide* to see how the concepts should be introduced. Read and discuss the page with the children. Provide concrete resources to support exploration.

1 Answer these.

Write 3 equivalent fractions.

a $\frac{1}{4}$

b $\frac{1}{2}$

c $\frac{1}{8}$

d $\frac{1}{10}$

Simplify.

e $\frac{9}{15}$

f $\frac{10}{15}$

g $\frac{12}{20}$

h $\frac{18}{24}$

Don't forget to simplify each fraction!

2 Answer these.

Write the equivalent fractions and decimals.

a 10%

b 50%

c 75%

d 20%

Write the equivalent fractions and percentages.

e 0.64

f 0.32

g 0.09

h 0.8

3 Measure.

Measure 1.2 litres of water into a container.

Write down these percentages of the volume in litres and then millilitres.

a 50%   b 25%   c 10%   d 20%   e 5%   f 1%

Now measure these volumes into different containers.

4 Investigate.

Egyptian fractions always had a numerator of 1. They were all unit fractions.

The Egyptians could add unit fractions to make another unit fraction. , e.g.:

$$\frac{1}{2} = \frac{1}{3} + \frac{1}{6}$$

Prove that this is correct.

Find 2 unit fractions that add up to make these:

$$\frac{1}{3} \quad \frac{1}{4} \quad \frac{1}{5}$$

What pattern can you see in the fractions? Does it continue?

Teacher's Guide

See page 33 of the *Teacher's Guide* for ideas of how to guide practice.
Work through each step together as a class to develop children's
conceptual understanding.

19 ★

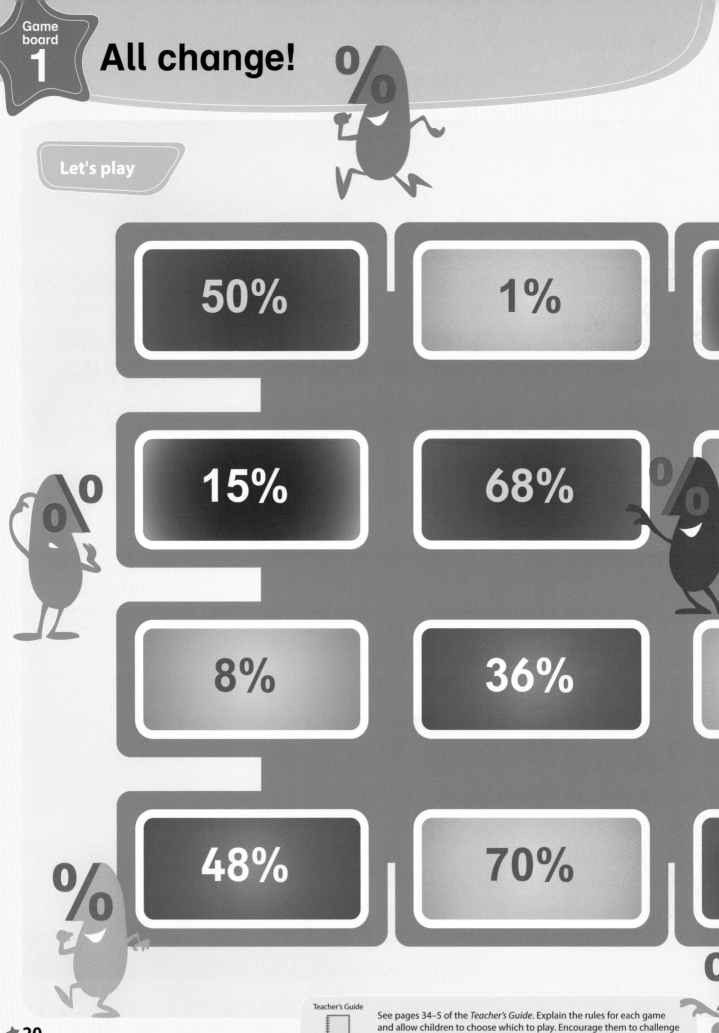

All change!

Let's play

50%	**1%**
15%	**68%**
8%	**36%**
48%	**70%**

Teacher's Guide

See pages 34–5 of the *Teacher's Guide*. Explain the rules for each game and allow children to choose which to play. Encourage them to challenge themselves and practise what they have learnt in the unit.

0%

75%

5%

24%

18%

55%

85%

56%

1 **Equivalence**

Choose a percentage. Write the equivalent fraction and decimal.

2 **Up the ladder**

Convert percentages to decimals. Arrange them in ascending order.

3 **Your game**

Design your own game using the gameboard. Explain the rules and play with a partner.

And finally ...

1

These 3 numbers are in descending order:
72 456.91, 34 822.6, 96 257

Explain where Eva has gone wrong and why.

2

Use the digits to make up a 3-digit number with 2 decimal places. Underneath your number, write what it would be after it is:

a   Multiplied by 10

b   Multiplied by 100

c   Divided by 10

d   Divided by 100

Round all your new numbers to the nearest whole number.

Repeat with another 3-digit number with 2 decimal places.

Teacher's Guide   See pages 36–7 of the *Teacher's Guide* for guidance on running each task. Observe children to identify those who have mastered concepts and those who require further consolidation.

★22

$\dfrac{3}{5}$ $\dfrac{2}{5}$ $\dfrac{7}{10}$ $\dfrac{1}{4}$

0.6 0.75

0.04 0.5

35% 8% 12% 60%

Pick 2 numbers from each of the clouds.

Write each number in 3 different ways, so that they are equivalent.

Each statement must include a fraction, a percentage and a decimal.

Order the fractions from largest to smallest. Explain how you did this.

Did you know?

We often use percentages to report survey results, because they are easy to compare.

Logitech, the computer accessory and remote controls manufacturer, found out that about 50% of lost remote controls were stuck between the cushions on a sofa, 4% were found in the fridge or freezer, and 2% turned up somewhere outdoors or in the car!

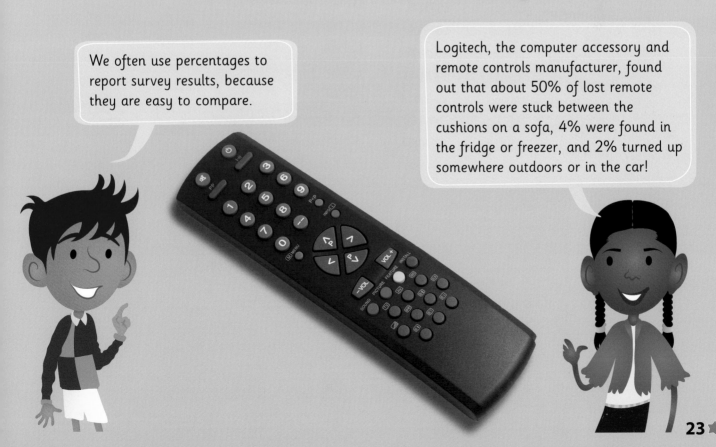

Calculations and algebra

Premier League Club	Season tickets Cheapest	Season tickets Most exp	Match-day tickets Cheapest	Match-day tickets Most exp	Cheapest day out	Programme
Arsenal	£1014.00	£2013.00	£27.00	£97.00	£36.30	£3.50
Aston Villa	£335.00	£615.00	£22.00	£45.00	£30.40	£3.00
Burnley	£329.00	£685.00	£35.00	£42.00	£42.30	£3.00
Chelsea	£750.00	£1250.00	£50.00	£87.00	£57.50	£3.00
Crystal Palace	£420.00	£720.00	£30.00	£40.00	£39.70	£3.50
Everton	£444.00	£719.00	£33.00	£47.00	£41.50	£3.00
Hull	£501.00	£574.00	£16.00	£50.00	£24.00	£3.00
Leicester	£365.00	£730.00	£19.00	£50.00	£27.50	£3.00
Liverpool	£710.00	£869.00	£37.00	£59.00	£45.80	£3.00
Man City	£299.00	£860.00	£37.00	£58.00	£45.80	£3.00
Man United	£532.00	£950.00	£36.00	£58.00	£45.50	£3.50
Newcastle	£525.00	£710.00	£15.00	£52.00	£23.30	£3.00
QPR	£499.00	£949.00	£25.00	£70.00	£33.40	£3.00
Southampton	£541.00	£853.00	£32.00	£52.00	£42.50	£4.00
Stoke	£344.00	£609.00	£25.00	£50.00	£33.60	£3.50
Sunderland	£400.00	£525.00	£25.00	£40.00	£33.20	£3.00
Swansea	£429.00	£499.00	£35.00	£45.00	£43.50	£3.00
Tottenham	£765.00	£1895.00	£32.00	£81.00	£41.00	£3.50
West Brom	£349.00	£449.00	£25.00	£39.00	£33.00	£3.00
West Ham	£620.00	£940.00	£20.00	£75.00	£29.00	£3.50

> Is the difference between the cheapest and most expensive season tickets always greater for London clubs than for the rest of the country?

Teacher's Guide
Look at the pictures with the children and discuss the questions.
See pages 38-9 of the *Teacher's Guide* for key ideas to draw out.

25 ⭐

Calculating mentally with 3- and 4-digit numbers

Let's learn

You need:
- number lines
- tape measures

I used rounding to help me make an estimate. My estimate for 1612 + 562 is 3000.

I don't think that is a very helpful estimate. You must think carefully about the rounding you use.

Solving problems mentally

Would you use addition or subtraction to solve these problems?

Can you solve all of these problems only using mental methods?

How much rainwater was collected between 2 hours and $2\frac{1}{2}$ hours?

A further 562.5 ml is collected between 4 hours and 5 hours. How much rainwater is there in the container now?

250 ml of water was collected after 1 hour. How much more water was collected after 2 hours?

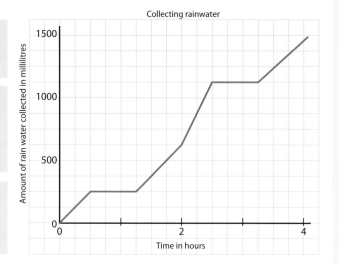

Making estimates

It is important to make an estimate before you calculate so that you have an idea of the size of the answer. This helps to avoid place-value errors, but the estimate has to be useful.

A useful estimate for the calculation 2344 + 1955 is 4000 or even 4300.

How have the numbers been rounded each time?

Now complete the calculation using the sequencing method and the rounding and adjusting method. How close is the estimate?

$$2344 + 1955 = 2344 + 1000 + 900 + 55$$
$$= 3344 + 900 + 55$$
$$= 4244 + 55$$

$$2344 + 1955 = 2344 + 2000 - 45$$

+ 2000

− 45

2344   4344

Teacher's Guide

Before working through the *Textbook*, study page 40 of the *Teacher's Guide* to see how the concepts should be introduced. Read and discuss the page with the children. Provide concrete resources to support exploration.

1 Solve.

Solve each calculation using a mental method of your choice.

a	4568 + 2980
b	1608 – 528
c	4847 + 1253

d	2003 – 1999 + 235
e	8467 – 208 – 1467
f	753.6 – 248 + 46.4

2 Create.

Create some addition and subtraction calculations using numbers from the grid, e.g. 823 – 725.

Make a useful estimate for each calculation.

Now decide and record which calculations you did using these mental methods.

6974	2052
9462	7409
8476	725
823	568

Rounding and adjusting	Partitioning and sequencing	Finding the difference
		823 – 725

Check your answers against your estimates.

3 Measure.

Measure the length and width of your table to the nearest millimetre.

Write the lengths in **millimetres (mm)**.

Find out:

a The difference between the length and width in millimetres.

b The perimeter of the table top.

c The length of 5 same-sized tables placed end to end.

Now measure other lengths in millimetres and make up a range of calculations.

4 Investigate

Solve this problem.

The cost of buying a total of 3 flight tickets with *Fly Airways* is £199 cheaper than buying 2 flight tickets to the same destination with *Star Choice Airline*.

No tickets cost less than £250.

Find as many ways as you can to make this true.

> Would the written method have been more efficient to answer the questions?

Teacher's Guide
See page 41 of the *Teacher's Guide* for ideas of how to guide practice. Work through each step together as a class to develop children's conceptual understanding.

27

Let's learn

I can calculate 90 + 50 × 100 in any order because addition and multiplication are both commutative operations.

You need:
- straws or lolly sticks
- ruler
- number lines

Yes, they are, but you can't do them in any order when they are in a calculation together. You have to use the order of operations.

Reordering calculations

Look at the calculations and how they have been re-ordered.

Is the answer still the same each time?

Why do you think it was useful to re-order the calculations?

14 − 23 + 11

14 (− 23) (+ 11)

14 (+ 11) (− 23)

35 × 8 ÷ 7

35 (× 8) (÷ 7)

35 (÷ 7) (× 8)

The order of operations

BIDMAS is a way to remember the order of operations. The table shows that division and multiplication have equal priority. So have addition and subtraction.

That is why you can move addition and subtraction or multiplication and division around in the examples above and the answers remain the same.

1st	**B**rackets
2nd	**I**ndices e.g. $(5 \times 2)^2$
3rd	**D**ivision & **M**ultiplication
Last	**A**ddition & **S**ubtraction

Ali's calculation is **90 + 50 × 100 =**

You can see from the table that you need to multiply first and then add.

Look at **(90 + 50) × 100**.

This time you need to add the numbers in the brackets first and then multiply the answer by 100. The answer is not the same.

Teacher's Guide

Before working through the *Textbook*, study page 42 of the *Teacher's Guide* to see how the concepts should be introduced. Read and discuss the page with the children. Provide concrete resources to support exploration.

1

Calculate.

Use BIDMAS to complete these calculations.

Think about re-ordering calculations when the priority is the same.

a $275 + 1850 - 175$

b $125 \div 15 \times 12$

c $4992 - 32 \times 6$

d 50×3^2

e $12 \times (79 + 21)$

f $48 \times 90 \div 30$

g $1800 \div (1 + 2^3)$

h $2\frac{3}{4} + (\frac{1}{3} \times 6) - \frac{1}{2}$

2

Solve.

These problems involve measurement. Think carefully about the units as well as the order of the operations.

a $\frac{1}{2}$ kg + 750 g × 3

b (3.75 m + 125 cm) ÷ 20

c (120 mm + 0.48 m) × 4

d Find 65 cm² more than the area of the square

11 cm

e James has $\frac{2}{5}$ litre of orange squash. Pete shares his 1000 ml jug of squash equally between himself and 3 other friends. How much more squash does James have than Pete? Write the calculation you used.

3

Measure.

Use straws or lolly stick to create different shapes.

Measure and compare the perimeters of 2 shapes each time, writing down the calculation you used.

Now calculate the difference.

Remember that you may need to use brackets to help you.

4

Investigate.

Choose some numbers and at least 2 operations from the grids.

Make up as many calculations as you can that give the answer 250.

Remember to think about BIDMAS.

800	30	1200	4
45	8	70	25
300	16	19	12
5	20	375	2

+	−	( )
×	÷	² or ³

Teacher's Guide

See page 43 of the *Teacher's Guide* for ideas of how to guide practice. Work through each step together as a class to develop children's conceptual understanding.

29 ★

Let's learn

That can't be right, your calculation has no numbers in it!

$a - b = c$

The letters represent values that we don't know yet, but we do know the relationship between them.

You need:
- number rods
- number shapes
- cubes

Understanding algebra

You can use number rods to help you explore algebra.

Look at the statement $b + c = a$

You add rod (b) and rod (c).
In total they equal rod a.

Now move the rods to represent each of the other statements.

a	
b	c

$b + c = a$ $a - c = b$
$c + b = a$ $a - b = c$

To find an unknown value in this example, you need to know the value of 2 of the rods or the relationship between them, e.g.

when $b = 3$ and $c = 6$, you can find the value of a because $b + c = a$.

Using algebra with four operations

Use number rods to create this representation.
What do you notice?

You can use different statements to express the relationship between the rods.

a			
c	d	d	d

$a = c + d + d + d$

You can write an algebraic expression to show the value of c.

You need to use the relationship between the rods to help you.

Rod c and 3 lots of rod d are equal to rod a, so you know that rod a subtract the d rods leaves rod c.

This can be written as $c = a - 3d$.

$a = c + (d \times 3)$

$a = c + 3d$

Teacher's Guide

Before working through the *Textbook*, study page 44 of the *Teacher's Guide* to see how the concepts should be introduced. Read and discuss the page with the children. Provide concrete resources to support exploration.

1 Write.

Use number rods to create these representations.
Write as many different algebraic expressions as possible each time.

a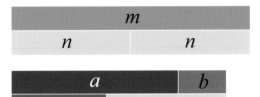

c

b

d

e Write some possible values for *m* and *n* in the first representation.

2 Match.

a 2 of the representations from Step 1 are represented below. Which ones?

 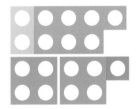 

b Write an algebraic equation for the representation that did not match any of the number rod representations in Step 1.

3 Solve.

Ali uses a formula to represent lengths of rope.

The total length of 2 pieces of rope a and 1 piece of rope b is equal to 3 pieces of rope c.

Use this information to write an algebraic equation.

Find possible lengths of rope *a* and *b* when rope *c* has the length 60 cm.

4 Think.

The value of rod *b* in each of these equations is 40.

a Use number rods and the representations shown here to help you find the values of *a* and *c*.

b Now make up some other values for *b* and record the values of *a* and *c* each time.

Teacher's Guide

See page 45 of the *Teacher's Guide* for ideas of how to guide practice. Work through each step together as a class to develop children's conceptual understanding.

31 ⭐

Dicey operations!

Let's play

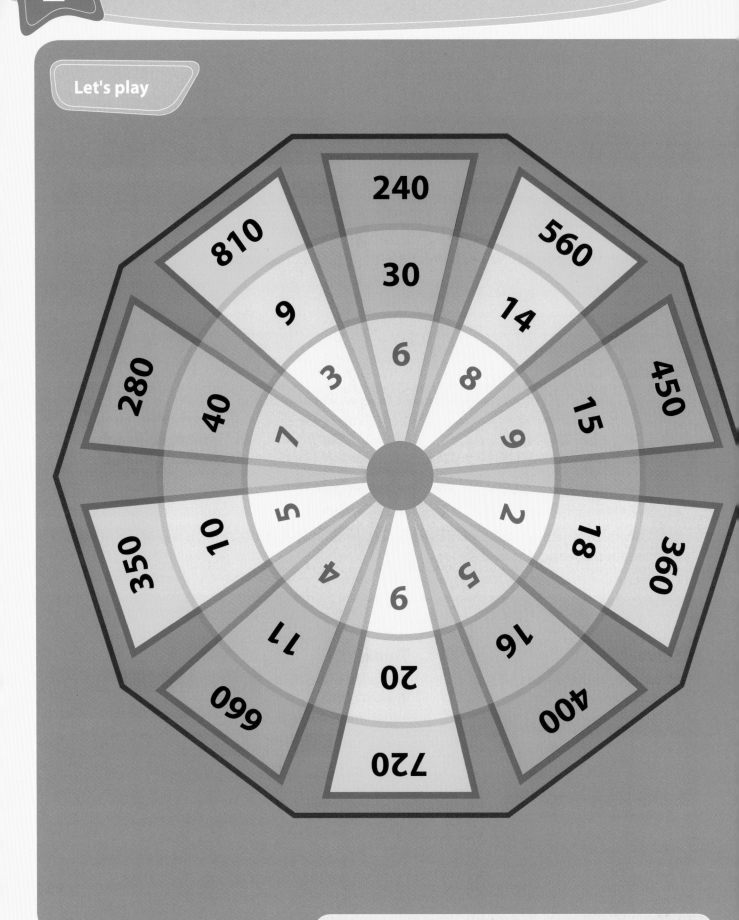

Teacher's Guide

See pages 46-7 of the *Teacher's Guide*. Explain the rules for each game and allow children to choose which to play. Encourage them to challenge themselves and practise what they have learnt in the unit.

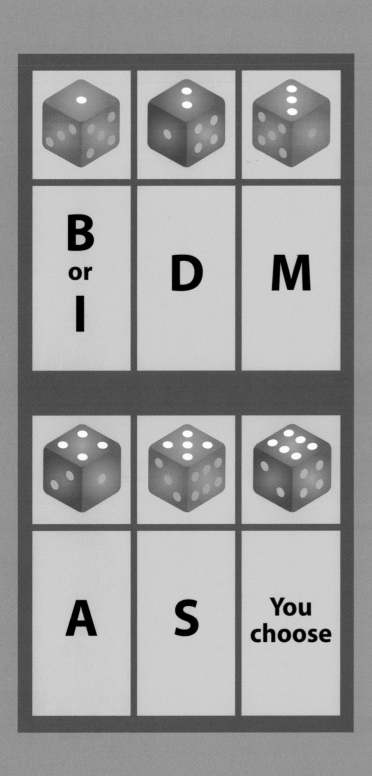

1 Aiming high

Choose a section from the wheel and roll the dice. Make a calculation with the largest possible answer.

2 Low score wins!

Play as for Aiming high but make the lowest answer.

3 Your game

Design your own game using the wheel. Explain the rules and play with a partner.

And finally ...

Let's review

1

Complete these calculations.

You need:
- place-value counters or Base 10 apparatus

a	$475 + 1199 =$	e	$3502 - 1199 =$
b	$477 + 1197 =$	f	$3505 - 1196 =$
c	$479 + 1195 =$	g	$3512 - 2199 =$
d	$481 + 1193 =$	h	$3525 - 1196 =$

What do you notice?

Explain your ideas using place-value counters to prove your thinking.

2

What do you think? Is Eva correct?

I think that all of these calculations give the same answer.

a $\quad 25 \times 4 - 1 =$ ☐

d $\quad 200 \div 4 + 7^2 =$ ☐

b $\quad$ ☐ $= 49 + 1000 \div 20$

e $\quad (2^2 + 5) \times (20 - 3^2) =$ ☐

c $\quad (320 + 76) \div 4 =$ ☐

f $\quad$ ☐ $= 6 + 120 - 3^3$

Write 4 different calculations of your own that all have the same answer. It can be a different answer to Eva's problem.

Make sure that each calculation uses more than 1 operation, e.g. × and +, and remember that you can use brackets, squares and cubes!

Teacher's Guide

See pages 48-9 of the *Teacher's Guide* for guidance on running each task. Observe children to identify those who have mastered concepts and those who require further consolidation.

3

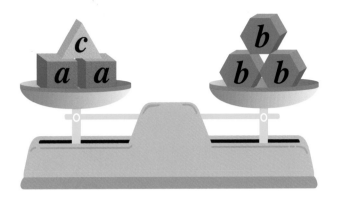

a   Write an algebraic expression to match the representation shown here.

b   Now find the mass of objects a, b and c using the information below:

$$a + c = 420\,\text{g}$$
$$c - a = 60\,\text{g}$$

Did you know?

Algebra was first developed in ancient Egypt and Babylon where people learned to solve many different equations.

The word *algebra* is Latin but comes from the Arabic word *al-jabr*. It was first used in the title of a book, *Hidab al-jabr wal-muqubala*, written in Iraq, in about 825 AD by the Arab mathematician Mohammad ibn-Musa al-Khwarizmi.

Larger numbers

How can I work out the cost of 8 of these if they cost £2.36 each?

DAFFODILS
£2.36
per bunch

I wonder how many slabs will be needed?

How much will the petrol cost?

I wonder if I have enough to make 50 cupcakes?

I wonder if there would be enough cordial to make drinks for everybody if we need 1 part of cordial to 4 parts of water?

1 litre

Teacher's Guide

Look at the pictures with the children and discuss the questions.
See pages 50–1 of the *Teacher's Guide* for key ideas to draw out.

37 ★

* whiteboard
* calculator

I've worked out 492 × 28 using long multiplication.

```
  492
×  28
 3936
  7 1
  984
 4920
 1 1 1
```

```
  492
×  28
 3936
   1
 9840
13776
   1
```

That can't be right. The answer is only 10 times bigger than the multiplicand. You missed out the 0 and multiplied by 2 instead of by 20.

Different methods of multiplication

The grid method shows what happens when you multiply 2 large numbers.

Look at 637 × 34.

The grid method

	600	30	7	
30	18 000	900	210	18 000 + 900 + 210 = 19 110
4	2400	120	28	2400 + 120 + 28 = 2548

You can shorten the writing by only partitioning the smaller number.

Short multiplication

```
   637      637     19110
×   30    ×   4    + 2548
 19110     2548     21658
  1 2      2 1 2      1
```

Long multiplication

```
    637
×    34
  19110
   1 2
   2548
   2 1 2
  21658
    1
```

You shorten the writing even more when you use this method.

Long multiplication in steps

multiplicand × multiplier = product

multiplicand → 8136 × 72 ← multiplier

Step 1

Multiply the multiplicand by the ones in the multiplier.

```
   8136
×    72
 16272
     1
```

Step 2

Multiply the multiplicand by the tens in the multiplier.

```
   8136
×    72
 16272
     1
569520
  2 4
```

Step 3

Add the 2 products to find the answer.

```
    8136
×     72
  16272
      1
+569520
   2 4
 585792
    1
```

Teacher's Guide

Before working through the *Textbook*, study page 52 of the *Teacher's Guide* to see how the concepts should be introduced. Read and discuss the page with the children. Provide concrete resources to support exploration.

1

Calculate.

Answer these using long multiplication.

a   76 × 17

c   7093 × 62

e   38 × 4801

b   423 × 28

d   92 × 639

f   59 × 98

2

Answer these.

Copy and complete these long multiplications.

a
```
        4 8
×     2 3
    ▢ ▢ ▢
    9 6 0
  1 1 0 ▢
```

b
```
      1 4 3
×       7 6
      8 5 8
  ▢ ▢ ▢ ▢ ▢
  1 0 8 6 8
```

c
```
        2 ▢ 1
×         5 ▢
      1 ▢ 2 7
  1 3 0 ▢ 0
  ▢ ▢ 8 7 ▢
```

3

Solve.

Solve these word problems using long multiplication.

a   One lap of a running track is 452 m. Dan runs 18 laps. How far has he run in kilometres?

b   A textbook weighs 368 g. A school buys 43 of them. How much do they weigh?

c   A brick is 38 cm long. Jane measures the length of the wall she plans to build. She discovers it will be 37 bricks long. How long is the wall?

d   It takes, on average, 23 minutes to make a paper bird using origami. How long does it take to make 1945 paper birds? Give your answer in days, hours and minutes.

4

Think.

a   Sam calculated 537 × 24 by using long multiplication. He got the answer 3222. How do you know the answer must be wrong? What was his error? How would you explain to him where and why he went wrong?

b   Sam has a 3-digit multiplier in the calculation 468 × 937. Work out the correct answer.

Check your answer with a calculator.

> Where do you need to write 0s in a long multiplication of two 3-digit numbers? Why?

Teacher's Guide

See page 53 of the *Teacher's Guide* for ideas of how to guide practice. Work through each step together as a class to develop children's conceptual understanding.

39 ★

Calculating mentally with large numbers

Let's learn

I can multiply 23 by 54 in my head. You multiply 20 and 50. That makes 1000. Then 3 × 4, which makes 12. Add them together and the answer's 1012.

That can't be right. 22 × 50 is 1100 and the correct answer must be more than that. You left out 4 × 20 and 3 × 50. You only multiplied the tens digits and the ones digits. 1242 is the correct answer.

Strategies for mental calculations

It is easy to multiply or divide by 10, 100 or 1000 in your head.

It can be easy to multiply and divide multiples of 10, 100 or 1000, but it depends on which multiple!

What is 400 × 30?

Multiply the 4 by 3 to get 12. Then multiply by 100 and by 10 to give 12 000.

This is what is happening mathematically:

$400 = 4 \times 100$ and $30 = 3 \times 10$

$$\begin{aligned}400 \times 30 &= 4 \times 100 \times 3 \times 10 \\ &= 4 \times 3 \times 100 \times 10 \\ &= 12\,000\end{aligned}$$

What is 5600 ÷ 80?

Divide the 56 by 8 to get 7 and the 100 by 10 to get 10 so the answer is 70.

This is what is happening mathematically:

$5600 = 56 \times 100$ and $80 = 8 \times 10$

$$\begin{aligned}5600 \div 80 &= 56 \times 100 \div (8 \times 10) \\ &= 56 \times 100 \div 8 \div 10 \\ &= 56 \div 8 \times 100 \div 10 \\ &= 70\end{aligned}$$

The operator, which is division, stays with the number that follows it, which is the divisor.

Deciding on a method

It is your choice how you do a calculation. You can do it all mentally, or write it all down. You could do a mixture of the two. Think about the most efficient way to do a calculation.

What is 600 × 25?

I can do 6 × 25 = 150 in my head because I can work out the 25 times table quickly. Then I multiply by 100 to get 15 000.

I can do 20 × 600 in my head and get 12 000. I need to jot that down before I work out 5 × 600 and get 3000. Then I can add on 12 000 to get the answer 15 000.

Both are quick ways to work out the answer.

Teacher's Guide

Before working through the *Textbook*, study page 54 of the *Teacher's Guide* to see how the concepts should be introduced. Read and discuss the page with the children. Provide concrete resources to support exploration.

1

Calculate.

Answer these:

a   200 × 40   c   510 × 20   e   101 × 30

b   200 ÷ 50   d   2500 ÷ 50   f   4020 ÷ 20

2

Calculate.

Decide how you would do each calculation mentally.
Explain your choice. Then answer the questions.

a   600 × 90   c   251 × 20   e   7820 × 24

b   3485 ÷ 17   d   2025 ÷ 25   f   2800 ÷ 20

3

Solve.

Solve these word problems mentally:

a   How many 20 pence coins do you need to make £12?

b   How many drinks of 150 ml can be served from 30 litres of juice?

c   Cheese is sold in 300 g portions. What is the weight, in kilograms, of 250 of the portions?

d   I have fifty 20 p coins and twenty 50 p coins. How much is that altogether?

4

Investigate.

a   There are 10 mm in 1 cm. Investigate how many square millimetres there are in 1 square centimetre. What about the number of square centimetres in a square metre?

b   What about volume? Explore the relationships between mm^3, cm^3 and m^3.

Teacher's Guide

See page 55 of the *Teacher's Guide* for ideas of how to guide practice. Work through each step together as a class to develop children's conceptual understanding.

41 ★

Multiply and divide up to 2 decimal places

Let's learn

I worked out the average times for the 100 m race on Sports Day. There were 8 runners and their times added up to 174 seconds. I got the answer 21.6 seconds.

That's not right. When you divide by 8 and the answer is a decimal it ends in 5. I think you made a mistake with the remainder. 21 × 8 = 168 so the remainder is 6. You carry on working out 6 ÷ 8 = 0.75 so the answer is 21.75 seconds.

Multiplying with decimals

You can use an array to help solve a multiplication calculation mentally.

Look at 7 × 0.3.

	1	1	1	1	1	1	1
0.1	0.1	0.1	0.1	0.1	0.1	0.1	0.1
0.1	0.1	0.1	0.1	0.1	0.1	0.1	0.1
0.1	0.1	0.1	0.1	0.1	0.1	0.1	0.1

7 is shown as 7 ones.

0.3 is shown as 3 tenths.

In total there are 21 tenths in the array.

$7 \times 0.3 = 21 \times 0.1$

$ = 2.1$

Dividing with decimals

You can use sharing to help solve a division calculation mentally.

Look at 0.24 ÷ 6.

1	1	1	1	1	1
0.01	0.01	0.01	0.01	0.01	0.01
0.01	0.01	0.01	0.01	0.01	0.01
0.01	0.01	0.01	0.01	0.01	0.01
0.01	0.01	0.01	0.01	0.01	0.01

0.24 is 24 lots of 0.01.

You can share 24 hundredths into 6 groups.

There are 4 hundredths in each group.

$4 \times 0.01 = 0.04$

So, $0.24 \div 6 = 0.04$

Calculating the mean

Here are the heights of 5 children: 1.22 m, 1.6 m, 1.38 m, 1.46 m, 1.54 m

You can find their average height. One type of average is called the mean.

The mean can be represented using bars.

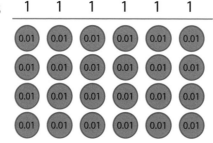

This shows the mean.

These show the heights.

To find the mean, first add all the heights:

$1.22 + 1.6 + 1.38 + 1.46 + 1.54 = 7.2$ m

Next, divide by the number of heights:

$7.2 \div 5 = 1.44$ m

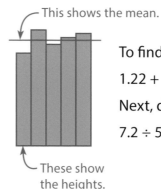

1

Calculate.

a   6 × 0.3   c   2.53 × 6   e   7.5 ÷ 4

b   9 × 1.24   d   82 ÷ 5   f   6.54 ÷ 3

2

Calculate.

Fill in the missing numbers.

a   0.2 × ▢ = 1.6   c   3 × ▢ = 0.8 × ▢   e   2.6 = ▢ ÷ 6

b   8.4 = 7 × ▢   d   8.75 ÷ ▢ = 1.75   f   9.2 = ▢ ÷ 5

3

Apply.

Carry out a survey to find the height of your classmates. Measure and check the height of each child in metres.

Find the mean height for the class.

Choose something else as the focus for another survey. Can you work out the mean from the responses you collect?

4

Think.

Roll a dice 4 times to generate 4 digits. Use them to fill in the spaces.

Aim to make the largest possible product. Now generate 4 digits to fit these spaces:

Aim to make the largest possible quotient.

Teacher's Guide

See page 57 of the *Teacher's Guide* for ideas of how to guide practice. Work through each step together as a class to develop children's conceptual understanding.

Solving problems with ratio and proportion

You need:
- number rods
- scales
- measuring jug
- sand
- water

Let's learn

I have 2 fish. One is 3 times as long as the other. The small one is 4 cm long and the longer one is 7 cm long.

That's not right. Your fish is only 3 cm longer than the other if it is 7 cm long. It should be 12 cm long if it is 3 times as long!

Multiplication as a relation

Compare the 2 number rods.

3 of the red rods fit against the green rod.

That means that the green rod is 3 times as long as the red rod.

3 is the scale factor that maps the length of the red rod to the length of the green rod

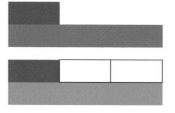

5 of the red rods fit against the orange rod.

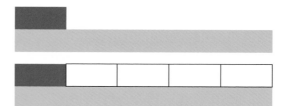

That means that the orange rod is 5 times as long as the red rod.

5 is the scale factor that maps the length of the red rod to the length of the orange rod.

Unequal sharing and grouping

Look at the bar.

You can use it to work out $\frac{3}{4}$ of 12.

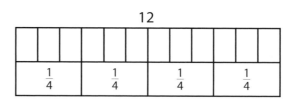

Work out $\frac{1}{4}$ by splitting 12 into 4 equal parts.

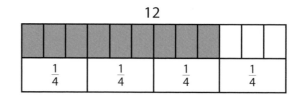

Each part is equal to 3. $\frac{3}{4}$ requires 3 of these.

$$\frac{3}{4} \text{ of } 12 = 3 \times 3$$
$$= 9$$

Teacher's Guide

Before working through the *Textbook*, study page 58 of the *Teacher's Guide* to see how the concepts should be introduced. Read and discuss the page with the children. Provide concrete resources to support exploration.

1 Calculate.

a   What number is 4 times as large as 9?

b   What number is 11 times as large as 12?

c   What number is 2 and a half times as large as 10?

d   What is the scale factor that multiplies 4 to get to 100?

e   What is the scale factor that multiplies 6 to get to 120?

f   What scale factor do you need to multiply 70 by to get 7700?

2 Answer these.

a $\frac{3}{4}$ of 28

b $\frac{2}{5} \times 35$

c $1\frac{1}{3} \times 24$

d   Write 24 as a fraction of 30 in its simplest form.

e   Write 45 as a fraction of 36 in its simplest form.

f   I am thinking of 2 numbers. One is $\frac{3}{4}$ of the other. One number is 12. What are the possible values of the other number?

3 Apply.

This cake recipe can be adjusted to make more or less cake, as needed.

200 g flour
75 g sugar
100 g butter
25 ml milk

Solve these problems using the recipe:

a   If I use 300 g of flour, how much butter should I use?

b   If I use 50 ml of milk, how much sugar should I use?

c   If I use 75 g of butter, how much flour should I use?

Weigh each new amount using sand and water.

4 Think.

I am thinking of a set of whole numbers.
The mean is $\frac{1}{5}$ of the total.
The total is 3 times the largest number.
2 of the numbers are $\frac{1}{6}$ of the total.
One of the numbers is the mean.
The smallest number is 4.
What are my numbers?

Teacher's Guide
See page 59 of the *Teacher's Guide* for ideas of how to guide practice. Work through each step together as a class to develop children's conceptual understanding.

45

Making products

Let's play

2 Is the answer between 3000 and 6000?

3 Is the answer a multiple of 5?

6 Is the answer a multiple of 3?

7 Is the answer less than 8000?

10 Will you work out 6 × 3 in your calculation? Justify!

11 Will you work out 10 × 10 in your calculation? Justify!

Teacher's Guide

See pages 60–1 of the *Teacher's Guide*. Explain the rules for each game and allow children to choose which to play. Encourage them to challenge themselves and practise what they have learnt in the unit.

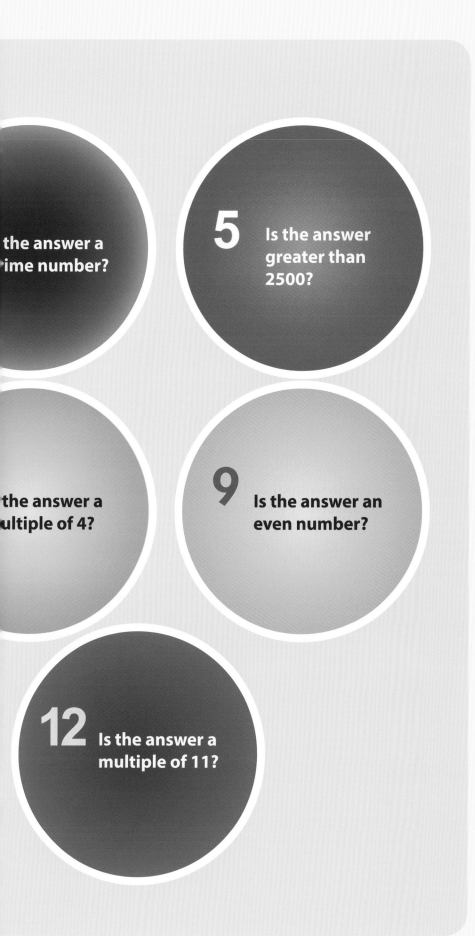

the answer a
rime number?

5 Is the answer
greater than
2500?

the answer a
ultiple of 4?

9 Is the answer an
even number?

12 Is the answer a
multiple of 11?

You need:

- two 1–6 dice
- calculator
- whiteboard

1 Predict

Roll the dice to make
two 2-digit numbers.

Roll again and answer
the question. Can you
work out the product?

2 Make it work

Roll the dice and add
the numbers to see
which question to
answer.

Roll the dice to make
2-digit numbers so
that the answer to the
question is 'yes'.

3 Your game

Design your own game.
Explain the rules and play
with a partner.

Let's review

1

Identify the error in each calculation. Write an explanation to the child who made the error so they learn where they went wrong. Write out the correct version.

a
```
      4 7 3
  ×     9 2
  ─────────
      8 4 6
  3 6 3 7 0
  ─────────
  3 7 2 1 6
```

b $8000 \div 20 = 40$

c $45.5 \div 7 = 6.3$

d 5 pencils cost 85p. How much do 8 pencils cost? Answer £6.80.

Can you identify where the child went wrong? Write down your explanation.

2

Work out each calculation using a mental calculation strategy and using long multiplication. Decide which method is more efficient in each case. Justify your answer.

a 565×28

b 380×49

Work out each calculation using a diagram and using a written method. Decide which method is more efficient in each case. Justify your answer.

c $\frac{3}{8} \times 48$

d $\frac{5}{6} \times 42$

Think about the actual numbers in your calculation to help you decide!

Teacher's Guide

See pages 62–3 of the *Teacher's Guide* for guidance on running each task. Observe children to identify those who have mastered concepts and those who require further consolidation.

3

Write a word problem for each calculation below. The pictures on page 36–7 may give you some ideas.

Work out the calculation.

a 847×78

b 600×80

c $28.6 \div 8$

d $\frac{3}{7} \times 63$

Did you know?

There are lots of different methods to multiply large numbers. Long multiplication, as you know it, was brought to Europe by the Arabic speaking people of Africa.

The lattice method, or gelosia, which is a different way of setting out long multiplication, originated in India in the 10th century and was brought to Europe in the 14th century by Leonardo of Pisa, also known as Fibonacci. Here is a gelosia, or lattice. Work out 849×54 using long multiplication and match where the numbers appear in each version. What other methods can you think of?

849×54

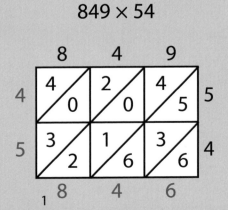

2-D shapes, 3-D shapes and nets

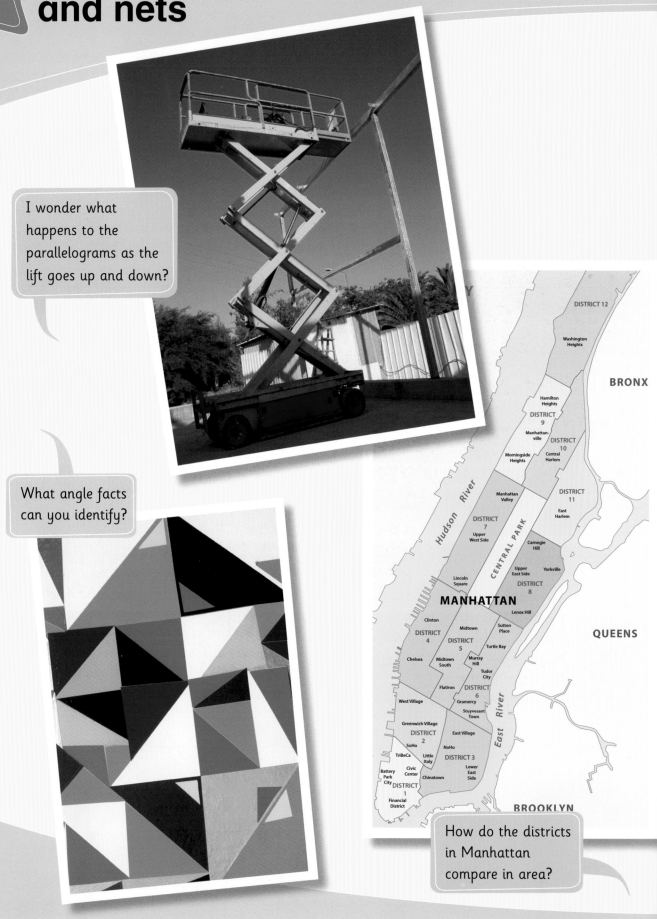

I wonder what happens to the parallelograms as the lift goes up and down?

What angle facts can you identify?

How do the districts in Manhattan compare in area?

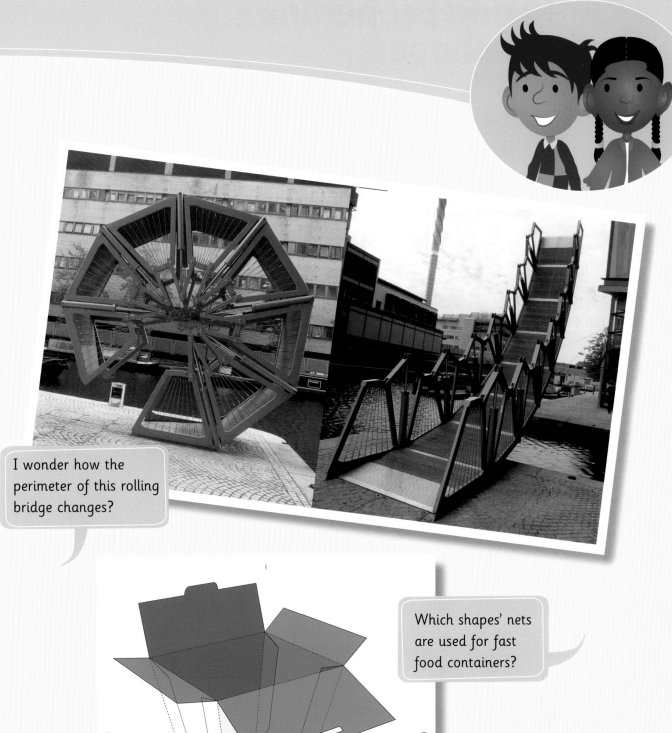

I wonder how the perimeter of this rolling bridge changes?

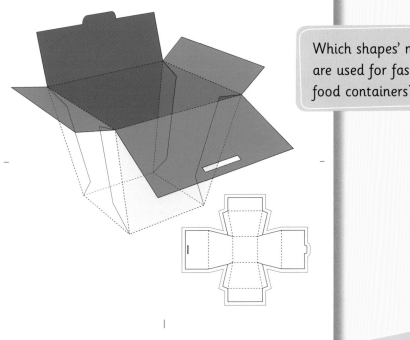

Which shapes' nets are used for fast food containers?

Teacher's Guide

Look at the pictures with the children and discuss the questions.
See pages 64–5 of the *Teacher's Guide* for key ideas to draw out.

51 ★

Area and properties of 2-D shapes

You need:
- 2-D shapes
- cm-squared paper
- protractor
- ruler
- tape measure

Let's learn

The area of a triangle is equal to the base multiplied by the height.

You're half right! It's equal to half the base multiplied by the height. If you don't divide by 2, you'll have the area of a parallelogram.

Finding the area of a triangle

1. The triangle has been flipped to make a parallelogram that is double its area.

2. Draw a perpendicular line.

3. Translate the small triangle over to the other side, to make a rectangle with the area = base × height.

This is twice the area of the original triangle, which explains why the formula for the area of a triangle is:

Area = $\frac{bh}{2}$

If $b = 10$ cm and $h = 6$ cm, then $A = \frac{10 \times 6}{2} = 30$ cm².

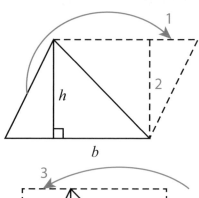

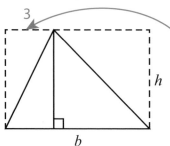

Finding the area of a parallelogram

In the first diagram, the parallelogram is made up from the original triangle and the dotted reflection.

The second diagram shows that the area of a parallelogram is equal to the area of a rectangle formed by translating the small triangle.
Area of parallelogram = base × the perpendicular height.

$A = b \times h$ or $A = bh$.

Remember you need the perpendicular height, not the length of the side.

If you know the area and the height of a parallelogram, you can find the length of its base,

e.g. $A = 60$ cm², $h = 6$ cm. Substituting into the formula $A = bh$ gives $60 = b \times 6$, so $b = 10$ cm.

Teacher's Guide

Before working through the *Textbook*, study page 66 of the *Teacher's Guide* to see how the concepts should be introduced. Read and discuss the page with the children. Provide concrete resources to support exploration.

1

Calculate.

Calculate the perimeter and area of these triangles.

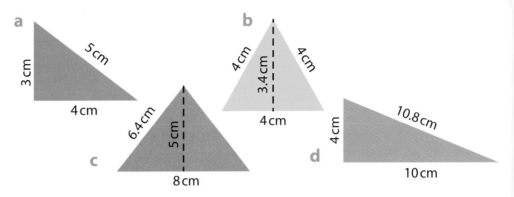

e What do you notice about the answers for a and b?

f What do you notice about the answers for c and d?

2

Calculate and draw.

Calculate the perimeter and area of the parallelograms.

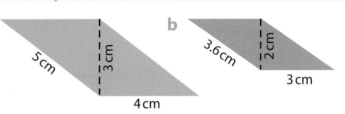

c Use cm² paper to draw diagrams of 2 parallelograms with an area of 16 cm².
Label the length of the base and the perpendicular height.

3

Apply.

The floor of your school hall has been damaged in a flood and needs replacing.

Measure the length and width of the hall to the nearest metre and find the area.

Estimate the cost of the alternative materials for replacing the floor and skirting board.

Check your calculations with a partner.

Skirting board

Solid oak	£9.50 per metre
Pine	£3.99 per metre

Flooring

Solid oak	£57.30 per square metre
Vinyl	£19.70 per square metre

4

Think.

a Investigate the area of triangles whose height is twice the length of their base.

b Can you find a pattern in your answers?

c Predict the area of a triangle with a base equal to x cm.

Teacher's Guide

See page 67 of the *Teacher's Guide* for ideas of how to guide practice. Work through each step together as a class to develop children's conceptual understanding.

4b Finding angles

You need:
- protractor
- ruler
- card
- hole punch
- paper fasteners

Let's learn

Vertically opposite angles are usually about the same size.

No, that's not quite right! Vertically opposite angles are *always* equal.

Vertically opposite angles

Vertically opposite angles are formed when 2 straight lines intersect. The lines create 2 pairs of equal angles.

Knowing that these angles are equal and that angles on a straight line total 180° will help you to solve missing angle problems.

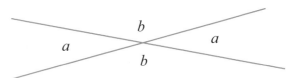

Angle $y = 130°$ because it is vertically opposite.

Angle $x = 180° - 130° = 50°$ because angle x and angle y make a straight line angle.

Angle z = angle $x = 50°$ because it is vertically opposite.

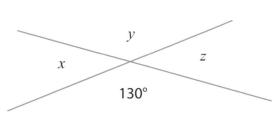

Equations with two unknowns

$x + y = 180°$ because they are angles on a straight line.

A good estimate for x is 120°. If $x = 120°$, then y must be 60°.

The 2 unknowns are linked. If $x = 117°$, then $y = 63°$, or if $x = 109°$, then $y = 71°$.

In every case, $x + y = 180°$.

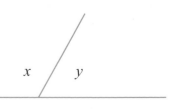

The exterior angle of a polygon is the angle between any side of a shape and a line extended from the next side. The sum of the exterior angle and the interior angle is 180° because it is a straight line.

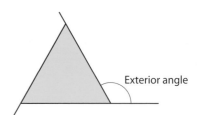

Exterior angle

Teacher's Guide

Before working through the *Textbook*, study page 68 of the *Teacher's Guide* to see how the concepts should be introduced. Read and discuss the page with the children. Provide concrete resources to support exploration.

1

Find solutions.

a Find possible values for x and y in degrees, when x is a multiple of 5° and is greater than 45° and less than 75°.

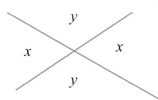

b Find possible values for a and b in degrees, when a is a multiple of 3° and is greater than 30° and less than 40°.

2

Unknown angles.

Work out the sizes of the unknown angles.

a

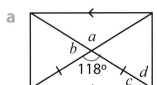

b

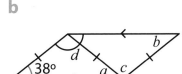

c

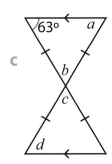

d

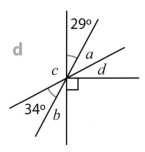

3

Measure.

Cut 2 pieces of card 12 cm by 1 cm and 2 pieces 6 cm by 1 cm.

Punch a hole 1 cm from each end of each piece of card.

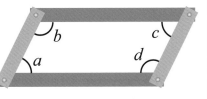

Fasten the pieces of card together with paper fasteners to make a parallelogram.

Measure the size of angles a, b, c and d.

Push the parallelogram to change the size of the angles and measure again. Repeat.

Use your results to describe the relationship of angles a, b, c and d.

4

Investigate.

a Draw 4 different regular polygons. Calculate the exterior angles. What do you notice?

b Do you think it would be the same for irregular polygons? Draw 2 irregular polygons and measure the exterior angles.

c Can you explain your findings?

Teacher's Guide

See page 69 of the *Teacher's Guide* for ideas of how to guide practice. Work through each step together as a class to develop children's conceptual understanding.

55

Describing 3-D shapes and making nets

You need:
- 3-D shapes
- 3-D shape construction kits
- cm-squared paper
- a food box

Let's learn

Any arrangement of 2 triangular faces and 3 rectangular faces will make a net that folds into a triangular prism.

I don't think so. There's more than 1 possible net, but not every combination folds to make a triangular prism.

Nets of a triangular prism

A net is an outline made when a 3-D shape is opened out flat.

Of these nets, only the first 3 fold to make a triangular prism.

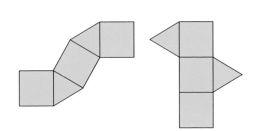
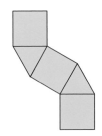
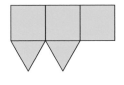

Formula for the volume of a cuboid

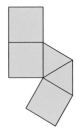

A cuboid is a 3-D shape. To find its volume you need 3 measurements: length, width and height.

The formula is:

Volume = length × width × height   or $V = l \times w \times h$ or simply $V = lwh$

The units are cubic units, e.g. cm³ (centimetres cubed), because you multiply cm × cm × cm.

The measurements can be multiplied in any order because multiplication is commutative.

If you know the volume, you can use the formula to find a missing measurement.

If $V = 36\,\text{cm}^3$, $l = 6\,\text{cm}$ and $w = 2\,\text{cm}$, you can substitute these values into the formula $V = lwh$:

$$36 = 6 \times 2 \times h$$
$$36 = 12h$$
$$h = 3\,\text{cm}$$

Teacher's Guide

Before working through the *Textbook*, study page 70 of the *Teacher's Guide* to see how the concepts should be introduced. Read and discuss the page with the children. Provide concrete resources to support exploration.

1 Answer these.

Which of these nets will make a 3-D shape? What shapes do they make?

a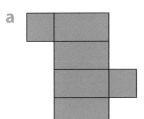

b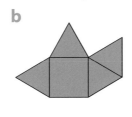

c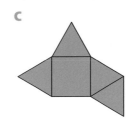

d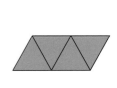

2 Answer these.

Find the volume of these cuboids.

a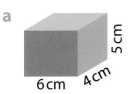

6 cm   4 cm   5 cm

b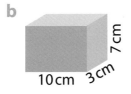

10 cm   3 cm   7 cm

c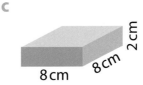

8 cm   8 cm   2 cm

d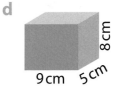

9 cm   5 cm   8 cm

e   Find 2 possible sets of measurements for a cuboid with a volume of 60 cm^3.

3 Measure and draw.

Find a food box that is a cuboid, e.g. a box of tea bags, biscuits or cereal.

Measure each side to the nearest cm.

Draw and label a small sketch of a possible net for the box.

Calculate the size of card that you would you need to make the net.

Draw an accurate net and make the cuboid.

4 Investigate.

Here is 1 net for a cube.

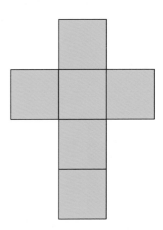

How many more nets can you find?

Teacher's Guide

See page 71 of the *Teacher's Guide* for ideas of how to guide practice. Work through each step together as a class to develop children's conceptual understanding.

57 ★

Area and volume snakes and ladders

Let's play

Finish	Move back **2** places		Move on **1** place		★

Teacher's Guide

See pages 72–3 of the *Teacher's Guide*. Explain the rules for each game and allow children to choose which to play. Encourage them to challenge themselves and practise what they have learnt in the unit.

Move back 3 places

Move on 3 places

Move on 2 places

⭐1 Area of triangle

Area	Move ...
1–4 cm²	1 place
4.5–12 cm²	2 places
12.5–18 cm²	3 places
> 18 cm²	4 places

⭐2 Volume of cuboid

Volume	Move ...
1–50 cm³	1 place
51–125 cm³	2 places
126–216 cm³	3 places
> 216 cm³	4 places
Any cube	Bonus 3 places

You need:

- 1–6 dice
- counters

 Triangle areas

Race to the Finish, creating triangles as you go. Calculate their area to work out your score. Who will be the winner?

 Cuboid volumes

Make cuboids and find their volume to work out your score. Move up the ladders but watch out for the snakes!

 Your game

Design your own game using the gameboard. Explain the rules and play with a partner.

And finally ...

1

Draw each diagram accurately on squared paper.

Explain which measurements you need to know to be able to find the area.

Now calculate the area for 3 of the shapes.

It is not possible to calculate one of the shapes accurately. Identify which one and explain why.

You need:
- squared paper
- ruler

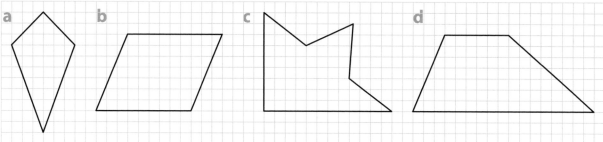

2

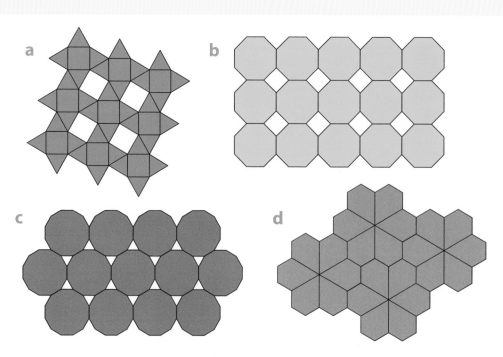

Without using a protractor, calculate the size of the angles in each pattern.
Explain how you found them.

Teacher's Guide

See pages 74–5 of the *Teacher's Guide* for guidance on running each task. Observe children to identify those who have mastered concepts and those who require further consolidation.

★**60**

3

These pictures appear on the 6 faces of a cube in a random order:

Here are 3 views of the cube:

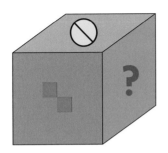 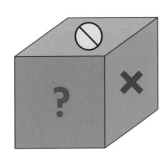 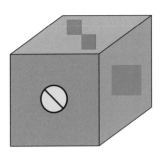

Draw a net for the cube, showing the pictures correctly placed.

Is your net correct?

Explain how you did it.

Did you know?

In 1985, scientists discovered a form of carbon in which 60 atoms were arranged to make a hollow sphere. They called it buckminsterfullerene. It's often called a 'buckyball'.

The scientists were surprised to discover that the structure was very similar to a football, composed of 20 hexagons and 12 pentagons, arranged in a regular pattern.

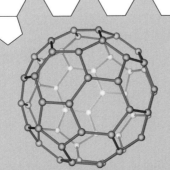

Numbers in everyday life

What are these?

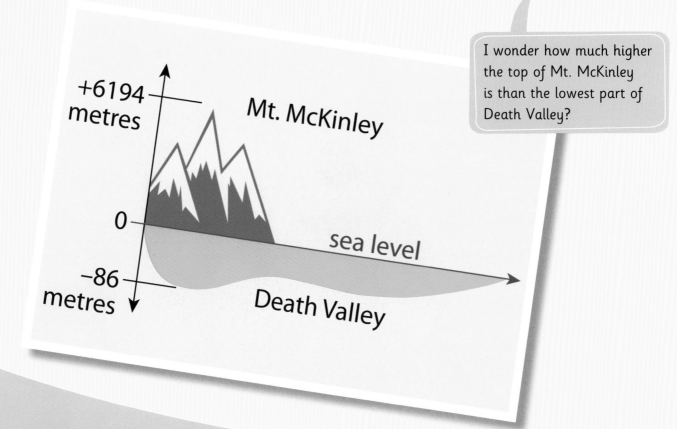

+6194 metres

Mt. McKinley

0

sea level

−86 metres

Death Valley

I wonder how much higher the top of Mt. McKinley is than the lowest part of Death Valley?

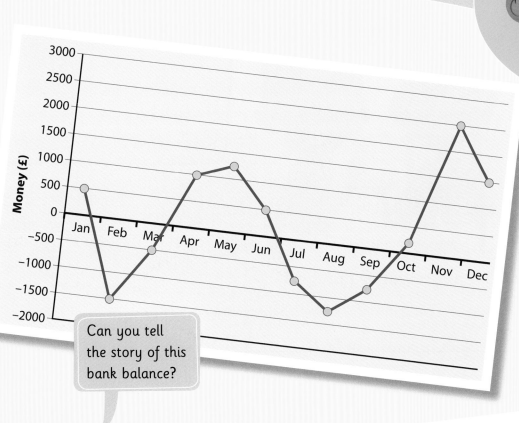

Can you tell the story of this bank balance?

What time is shown on this stopwatch?

Teacher's Guide

Look at the pictures with the children and discuss the questions.
See pages 76–7 of the *Teacher's Guide* for key ideas to draw out.

63 ⭐

Let's learn

You need:
- counters (different colours)
- temperature and time graphs
- bank statements
- squared paper
- number line

6 is bigger than 2, so if we add −6 and 2, the answer will be bigger than 2.

I don't think that is correct, because −6 is smaller than 2. If you count on 2 from −6 you get −4.

Calculating with positive and negative numbers

Adding a positive number ⟶   ⟵ Subtracting a positive number

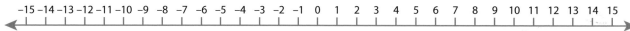

⟵ Adding a negative number   Subtracting a negative number ⟶

The blue counters represent negative numbers.
The red counters represent positive numbers.

If you add the red counters to the blue counters you are left with −1:

$-4 + 3 = -1$

If you add the blue counters to the red counters you are also left with ⁻1:

$3 + -4 = -1$

Why are the answers the same?

Negative numbers on graphs

You use negative numbers for coordinates, on line graphs and maps.

This grid has 4 quadrants and uses negative numbers as well as positive numbers.

The first number in a coordinate pair tells you the x-position. The second number tells you the y-position.

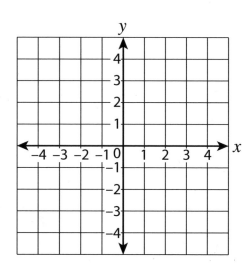

Teacher's Guide

Before working through the *Textbook*, study page 78 of the *Teacher's Guide* to see how the concepts should be introduced. Read and discuss the page with the children. Provide concrete resources to support exploration.

1

Answer these.

a   −16 + 30

b   −24 + 10

c   24 − 50

d   33 − 45

e   12 + −24

f   5 − −11

g   −26 − −13

h   −42 + −29

2

Plot.

On squared paper, draw a coordinate grid with 4 quadrants, like the one opposite.

Mark these coordinates with a cross:

(1, 4), (−5, −3), (3, −2), (−4, −6), (5, 2), (−5, −2), (5, −4), (−1, 5)

Join the crosses to make a 2-D shape.

What shape did you make?

3

Apply.

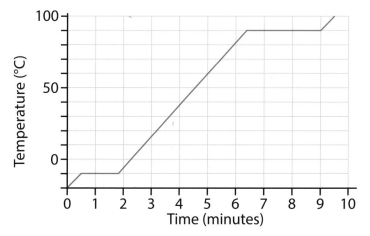

What is the difference between the highest and lowest temperatures?

Make up a story to show what is happening.

Draw your own line graph and tell its time and temperature story.

4

Think.

Answer these calculations:

4 + 2 and 4 + −2

4 − 2 and 4 − −2

What do you notice about the answers?

Teacher's Guide

See page 79 of the *Teacher's Guide* for ideas of how to guide practice.
Work through each step together as a class to develop children's conceptual understanding.

65 ★

5b | Decimals in context

You need:
- place-value grid
- digit cards
- stopwatch
- jugs
- scales
- sand

Let's learn

1 litre is the same as 100 millilitres. All we do to convert from litres to millilitres is multiply by 100.

I disagree. To convert litres to millilitres we multiply by 1000, because there are 1000 millilitres in 1 litre. It's the same for grams and kilograms.

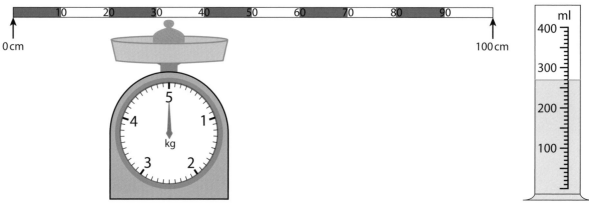

Two decimal places in context

100 cm are equivalent to 1 m. So 1 cm is $\frac{1}{100}$ of a metre.

You can write 1 m 63 cm as 1.63 m, because 63 cm is $\frac{63}{100}$ of a metre.

To convert metres to centimetres, multiply by 100.

To convert centimetres to metres, divide by 100.

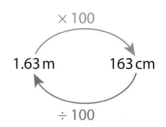

Three decimal places in context

1000 g are equivalent to 1 kg. So 1 g is $\frac{1}{1000}$ of a kilogram.

You can write 2 kg 125 g as 2.125 kg, because 125 g is $\frac{125}{1000}$ of a kilogram.

To convert kilograms to grams, multiply by 1000.

To convert grams to kilograms, divide by 1000.

1000 ml are equivalent to 1 litre. So 1 ml is $\frac{1}{1000}$ of a litre.

Can you write 1 litre 725 ml in 2 different ways?

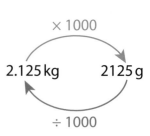

Teacher's Guide

Before working through the *Textbook*, study page 80 of the *Teacher's Guide* to see how the concepts should be introduced. Read and discuss the page with the children. Provide concrete resources to support exploration.

1

Answer these.

Write these measurements in metres with 2 decimal places:

a   3 m 25 cm

b   12 m 17 cm

c   5756 cm

d   2708 cm

Write these measurements in litres with 3 decimal places:

e   1 l 347 ml

f   10 l 903 ml

g   7870 ml

h   9363 ml

2

Answer these.

Write these times in seconds to 2 decimal places.

a   0:02:15.65

b   0:05:24.38

c   0:06:13.12

d   0:10:15.75

e   0:12:10.36

f   0:10:54.82

3

Apply.

Think of 4 small tasks, e.g.:

• building a tower of 20 interlocking cubes

• writing the numbers 1 to 50.

Work with a partner.

Use a stopwatch to time each other as you do the tasks.

Record the results in seconds to 2 decimal places.

4

Think.

Use these digits to make as many different gram masses as you can. You need to use all 4 digits for each mass.

Be systematic!

Once you have made them, write them in 2 other ways.

6   3   9   2

Teacher's Guide

See page 81 of the *Teacher's Guide* for ideas of how to guide practice. Work through each step together as a class to develop children's conceptual understanding.

Number darts

Let's play

Teacher's Guide
See pages 82–3 of the *Teacher's Guide*. Explain the rules for each game and allow children to choose which to play. Encourage them to challenge themselves and practise what they have learnt in the unit.

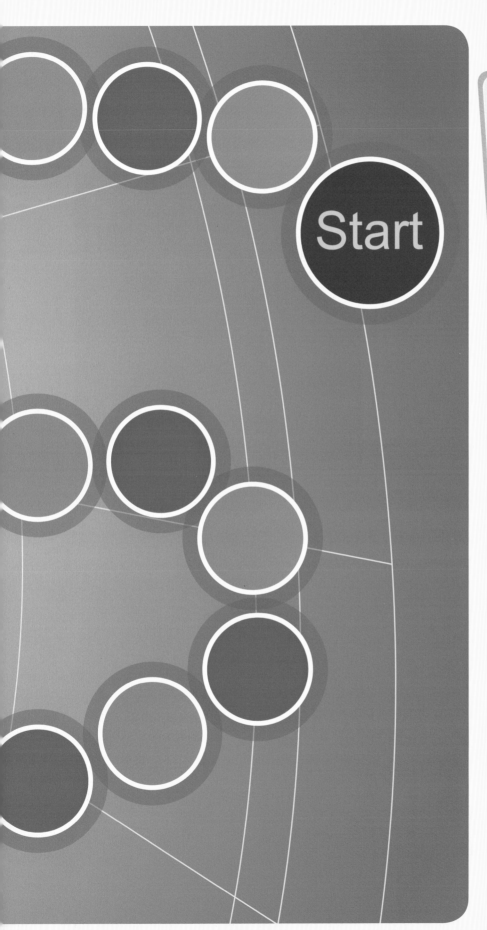

Start

You need:

- 2 decimal point cards
- 2 different coloured counters
- 2 sets of digit cards
- number cards from 10 to −10 (not 0)
- 1–6 dice

1 **How low can you go?**

Make the lowest number with 3 decimal places to move along the track.

2 **Add or subtract?**

Add and subtract positive and negative numbers to move along the track.

3 **Your game**

Design your own game using the gameboard. Explain the rules and play with a partner.

And finally ...

1

Anya is looking at a drawing of a coastal stretch of land.

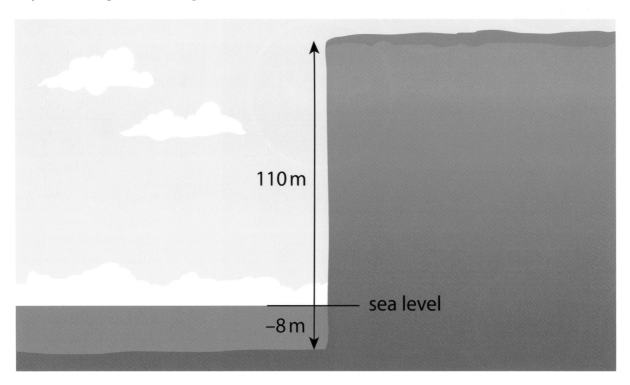

110 m

sea level

−8 m

It shows the height of the cliff and the depth of the water.

Anya thinks that the difference between them is 102 m.

What is it really?

Explain why Anya is wrong.

2

Find the equivalent measure that has decimal places.

a   1 kg 357 g   e   3 kg 400 g   i   2 kg 50 g

b   2 l 456 ml   f   1 l 300 ml   j   3 m 3 cm

c   5 m 36 cm   g   6 m 10 cm

d   5 cm 4 mm   h   1 l 75 ml

Teacher's Guide

See pages 84–5 of the *Teacher's Guide* for guidance on running each task.
Observe children to identify those who have mastered concepts and those who
require further consolidation.

3

| 2 | 5 | 8 | 6 | · |

Use the digits to make up some volumes in litres that have 1, 2 and 3 decimal places.

Find all the possibilities.
Explain how you know you have found them all.

Write each volume using litres and millilitres.

> For example,
> 28.86 l = 28 l 860 ml

Did you know?

Negative numbers weren't commonly used in Europe until the nineteenth century.

In India negative numbers were used much earlier. The mathematician Brahmagupta wrote the first set of rules in the seventh century. He used the ideas of fortunes and debts for positive and negative.

Solving problems

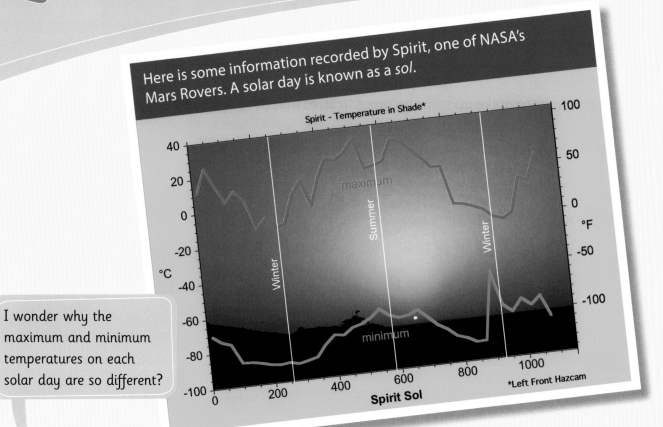

Here is some information recorded by Spirit, one of NASA's Mars Rovers. A solar day is known as a *sol*.

I wonder why the maximum and minimum temperatures on each solar day are so different?

I think there is approximately £200 in this picture. What do you think?

Ali said he completed his homework in a fraction of the time that it took me.
I wonder how long it took Ali?

Eva's start time

Eva's finish time

What will the next shape in the sequence look like?

Teacher's Guide
Look at the pictures with the children and discuss the questions.
See pages 86–7 of the *Teacher's Guide* for key ideas to draw out.

73 ★

Let's learn

You need:
- bead strings
- + and −
  number lines

When I count on or back in tens, the ones digit never changes. So 7 − 10 = −7.

That's true most of the time, but not when you have to cross zero. We could use number bonds to 10 to help us!

Calculating intervals and crossing zero

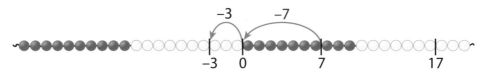

The calculation 7 − 10 = [ ] can be rewritten as 7 − 7 − 3 = [ ] to help cross zero.

The first number subtracted gives zero. Then you can use number bonds of 10 to identify how much more needs to be subtracted.

In the same way, 53 − 100 = [ ] can be re-written as 53 − 53 − 47 = [ ] using the number bonds of 100 to help you.

Explain to your partner where you can see 53 − 100 using the the subtraction sequence 53 − 53 − 47.

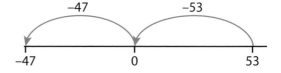

Solving problems involving addition and subtraction

The bead string or a number line can be used to help represent this problem.

How much colder is the temperature in the freezer than in the fridge?

Freezer   Fridge

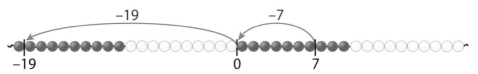

Here, you can see that the difference between 7 °C and −19 °C is the interval made up of 7 beads and 19 beads. The difference is 26 °C.

Teacher's Guide

Before working through the *Textbook*, study page 88 of the *Teacher's Guide* to see how the concepts should be introduced. Read and discuss the page with the children. Provide concrete resources to support exploration.

1

Solve.

Use a bead string and a number line to help you.

a   18 – 20 = ☐

b   18 – 30 = ☐

c   The difference between –9 and 15 is ☐ .

d   Count on 22 from –5 and land on ☐ .

e   Count back 18 from 6 and land on ☐ .

2

Calculate.

a   89 – 100 = ☐

b   75 – 77 = ☐

c   175 – 177 = ☐

d   275 – 377 = ☐

e   750 – 1000 = ☐

f   –25 + ☐ = 50

> Remember to think about number bonds!

3

Compare.

a   Use the line graph on page 72 to calculate the difference between the minimum and maximum temperature on Mars on the 400th solar day.

b   Now compare other maximum and minimum temperatures in the same way. Record your findings in a table.

4

Think.

Copy and complete the grid so that the value in the centre square is the difference between the pairs of numbers on either side of it.

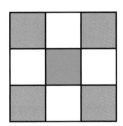

–27	74	147
120	–73	–47
75	–72	100

Is it possible to find a solution where the sum of all orange corners is a single digit?

Teacher's Guide

See page 89 of the *Teacher's Guide* for ideas of how to guide practice. Work through each step together as a class to develop children's conceptual understanding.

75 ★

6b Solving multi-step problems

Let's learn

When you see the words 'more than' in a problem, you always need to use addition.

That's not always true. Sometimes you need to subtract. It depends on how the words are used in the sentence!

You need:
- coins 5p 1p 10p
- large squared paper

Multi-step money problems

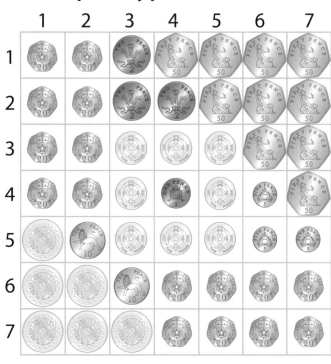

Make the money grid using the coins shown.

First make an estimate of how much money there is in the grid altogether.

Discuss the strategies you can use to find the total value of the grid.

Use additive reasoning: £2.00 + £2.00 + £2.00 + £2.00 + £2.00 + £2.00

Use multiplicative reasoning: £2.00 × 6

Choosing operations

Now use the grid of coins and look at the following problems. You need to decide how many steps there are to the problem and which operation to use each time.

- *Which row has £3.26 more than row 2?*
- *Column 3 has £2.69 more than column*
- *The difference between two 2 by 2 squares is 70p. Find a way to make this true.*

?	?
?	?

and

?	?
?	?

2 of the problems use the words 'more than' but you have to use different operations to solve them.

Teacher's Guide

Before working through the *Textbook*, study page 90 of the *Teacher's Guide* to see how the concepts should be introduced. Read and discuss the page with the children. Provide concrete resources to support exploration.

1

Calculate.

a   £3.79 + £4.55 – £1.50 =

b   £10.40 – £4.99 + £5.10 =

c   £5.28 more than £12.72 is ⬚

d   How much more is £20 than the sum of £2.78 and £9.45?

e   The total of £15.28 and £3.70 is £9.99 more than ⬚

2

Answer these.

a   Calculate the sum of any 2 of the grid sections below, a group of 3 of the grid sections and then all 4 grid sections.
Write the calculations you use each time.

1    **2**   **3**    **4**

b   Now find the differences between any 2 of the grid sections. You should find 6 differences.

3

Solve.

Eva decides to remove the coins from 1 column in the money grid on page 76.

a   Which column should she remove to leave the greatest amount of money on the grid? What if she leaves the smallest amount of money? How much money is left each time?

b   How do the amounts change if she removes 1 row instead of 1 column?

4

Think.

 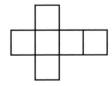

Place these 2 shapes anywhere on the grid. You can rotate the shapes if you like. Remove the coins that are covered. Calculate how much money is left on the grid.

Now explore different ways to use both shapes to investigate this statement:

The total money left on the grid each time is always greater than the sum of the coins removed.

Prove your findings.

Teacher's Guide

See page 91 of the *Teacher's Guide* for ideas of how to guide practice. Work through each step together as a class to develop children's conceptual understanding.

77 ★

Rounding to solve problems

Let's learn

I must round the answer to my word problem up to 4 km because the digit 5 is after the decimal point.

3.525 km

Not necessarily, you must check what the question is asking!

Rounding

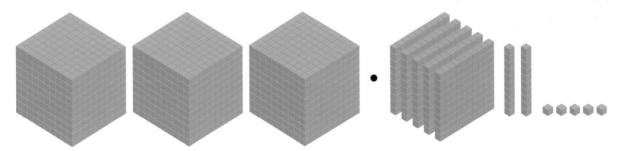

It is important to know the place value of the digits in a number before rounding.

Consider the positional and multiplicative aspects of the number represented above too.

Now you have the values of all the parts of the number you add them together to get the whole number:

$$3 + \frac{5}{10} + \frac{2}{100} + \frac{5}{1000} = 3.525$$

You can use the Base 10 apparatus to show how the number is rounded to 2 decimal places.

3.525 rounds to 3.53 because the $\frac{5}{1000}$ rounds up to the next $\frac{1}{100}$.

Think about how this changes when you round to only 1 decimal place.

Rounding to solve problems

Look at Eva's word problem below:

Hayley is taking part in a 20 km cycle race. She has completed 16.475 km so far.
How many whole kilometres does she have left to cycle?

You will need to use rounding here because the question asks you to think about whole kilometres and the distance given includes decimals.

Eva's answer is incorrect because she simply used what she knows about the rules of rounding.

Discuss how the question can be changed so that Eva's answer is correct.

Teacher's Guide

Before working through the *Textbook*, study page 92 of the *Teacher's Guide* to see how the concepts should be introduced. Read and discuss the page with the children. Provide concrete resources to support exploration.

1

Answer these.

a   Explain the positional and multiplicative aspects of each number to your partner.

4.054   5.184   **10.49**   12.305   **0.841**   0.006

b   Round all the blue numbers to the nearest tenth and all the yellow numbers
to the nearest hundredth.

2

Convert.

Here are the distances of
3 hiking routes in Scotland.

$1\frac{3}{4}$ km $2\frac{5}{8}$ km $2\frac{4}{5}$ km

Blue   Yellow   Orange

a   Write the decimal equivalent of each route.

b   Using the decimal equivalents, find the sum of and difference between all pairs of routes.

c   Round each answer to the nearest tenth of a kilometre.

3

Apply.

Use a stopwatch to time your group doing
different activities, e.g. counting in sevens to 100
or writing the alphabet backwards.

a   Record the time taken for each activity as,
e.g. 1:46 14, to include fractions of a second.

b   Order and compare
the group's times,
rounding differences
calculated to the
nearest **whole**
second.

4

Think.

Eva adds together pairs of
decimal numbers:

5.   and **8.**
Her sum **rounds** to 13.2 each
time. No sum is exactly 13.2

a   Find at least 5 ways to
make this true.

b   Which pair of numbers to
2-decimal places can Eva
use so that both numbers
have the smallest
possible value?

Teacher's Guide

See page 93 of the *Teacher's Guide* for ideas of how to guide practice.
Work through each step together as a class to develop children's
conceptual understanding.

79 ★

You need:
- cubes, counters or pegs
- number rods

Let's learn

You always need to know the previous term in a linear sequence before you can work out the next term.

It's useful to look at previous terms and next terms to help you find a rule, but you don't need to know the previous term each time once you have found the rule.

Describing patterns and linear sequences

When working with sequences, it is important to look closely at the pattern and decide what is the same and what is different each time.

Here, the number of red counters and blue counters increase by 1 each time, but there is always 1 more red than blue.

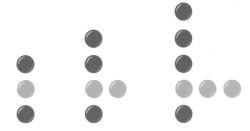

You can use what you know to help you find future terms in the sequence.

Using simple formulae

Tables are a good way of spotting patterns.

	1st term	2nd term	3rd term	4th term
●	1	2	3	4
●	2	3	4	5
Total	3	5	7	9

The number of blue counters can be described simply as the **term (or position)**.

The number of red counters in each term can be described as the **term (or position) plus 1**.

You can then use an algebraic expression using n to represent the position of the term each time. So the number of red counters can be expressed as $n + 1$.

However the total number of counters each time has a different pattern.

You can express the total number as **next term = previous term + 2 or Total = $2n + 1$**.

Teacher's Guide

Before working through the *Textbook*, study page 94 of the *Teacher's Guide* to see how the concepts should be introduced. Read and discuss the page with the children. Provide concrete resources to support exploration.

1

Calculate.

Use the formula $n + 1$. Calculate the total number of red counters each time as the pattern on page 80 continues.

a   5th term   c   10th term   e   25th term   g   200th term

b   7th term   d   20th term   f   100th term   h   1000th term

2

Calculate

Use the formula $2n + 1$. Calculate the **total** number of counters each time as the pattern on page 80 continues.

a   7th term   d   24th term

b   10th term   e   75th term

c   16th term   f   Which term in the sequence will have a
                                                   total of 101 counters?

3

Apply.

Use the algebraic rule: Total = $3n + 2$.

Use counters or cubes to make up a pattern to follow the rule.

a   Record the total number in
     each term in a table.

b   Predict the total number of
     counters in the 100th term,
     and other terms of your choice.

4

Think.

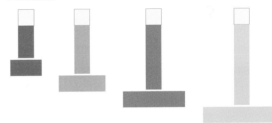

Use number rods to represent the start of the linear sequence shown here.

The white cube has the value 1, the red bar has the value 2 and so on.

a   Investigate to find the total value of
     each term in the sequence and find
     a way to describe the rule in words.

b   Express the rule algebraically using
     n and use it to find the value of
     other terms, e.g. the 50th term.

Teacher's Guide

See page 95 of the *Teacher's Guide* for ideas of how to guide practice.
Work through each step together as a class to develop children's
conceptual understanding.

81 ★

Formula won!

Let's play

–1	7	14	18	8	21
11	19	5	11	13	12.5
20	13	11	45	–2	15
12	23	17	27	39	3
15	13	10.5	8	16	12
17	3	11.5	14	15	33

Teacher's Guide
See pages 96–7 of the *Teacher's Guide*. Explain the rules for each game and allow children to choose which to play. Encourage them to challenge themselves and practise what they have learnt in the unit.

Four in a row grid

$4n - 5$	$3n + 2$	$5n - 7$
$n + 11$	$(n \div 2) + 10$	$6n + 9$

1 Four in a row

Roll the dice and choose a formula. You want to be the first player to place 4 counters in a row on the grid, either horizontally, vertically or diagonally.

2 Making squares

Roll the dice and use a formula to cover squares of 4 numbers on the grid, e.g. numbers 11, 45, 17 and 27.

3 Your game

Design your own game using the gameboard. Explain the rules and play with a partner.

Making squares grid

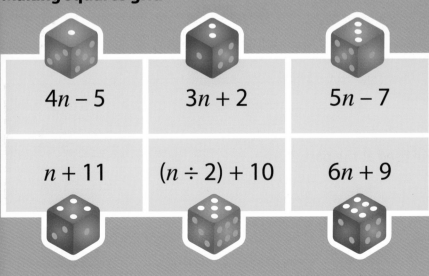

$4n - 5$	$3n + 2$	$5n - 7$
$n + 11$	$(n \div 2) + 10$	$6n + 9$

Let's review

1

Complete these sequences of numbers.

a   4750, 3500, ⬚ , 1000, ⬚ , ⬚

b   ⬚ , −12, ⬚ , 60, 96, ⬚

c $-\frac{9}{8}$, $-\frac{3}{4}$, ⬚ , 0, ⬚ , ⬚ , ⬚

d   Fill in the missing numbers on the number line.

−225   150   300

2

The pie chart shows the number of millitres of different drinks that children drank at a party.

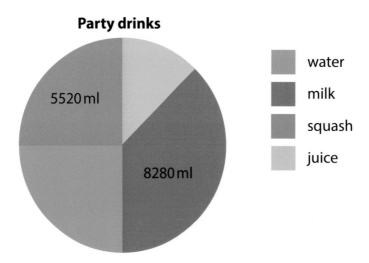

Party drinks

5520 ml

8280 ml

- water
- milk
- squash
- juice

Juice is sold in 1 litre cartons.

How many **whole** cartons of juice were used at the party?

3 Continue this sequence of cubes.

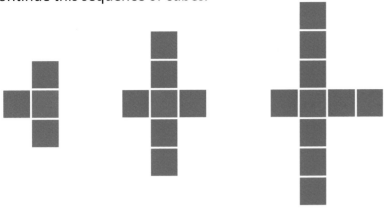

a Describe how the sequence changes each time.

b Draw a table to show the number of cubes in each term.

c Write an algebraic expression using n to describe the sequence.

d Predict some future terms in the sequence.

Did you know?

The Fibonacci Sequence is the series of numbers: **0, 1, 1, 2, 3, 5, 8, 13, 21, 34, ...**
The next number is found by adding the 2 numbers before it.

Fibonacci (or Leonardo Pisano) lived between 1170 and 1250 in Italy. He was the son of Guglielmo Bonacci. 'Fibonacci' is a shortened name for 'Filius Bonacci' translated to mean 'son of Bonacci, Among the great contributions that he made to mathematics, the Fibonacci sequence is one of the most famous. However, Fibonacci was not the first to know about the sequence, it was known in India hundreds of years before! The sequence was used to solve the problem, 'How many pairs of rabbits are created by 1 pair in 1 year?'

When squares are drawn whose sides are the same length as the numbers in the Fibonacci sequence, the pattern created is a spiral. This is known as the Fibonacci spiral.

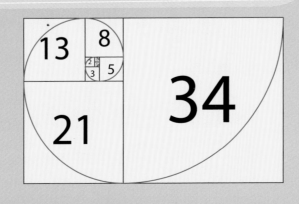

Let's explore fractions and algebra!

What fractions can you see in this picture?

How could you show this information on a pie chart?

How else could you represent 113.5?

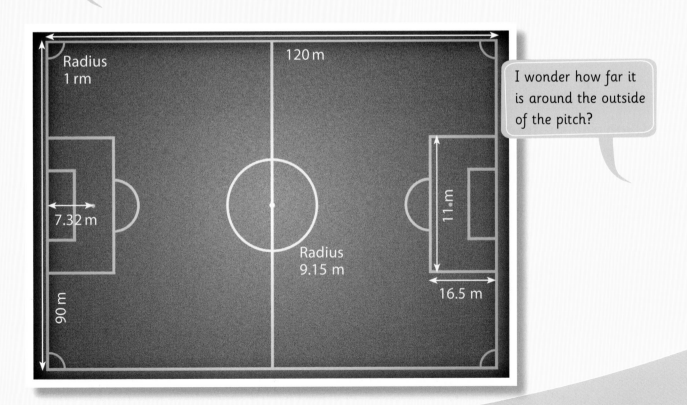

I wonder how far it is around the outside of the pitch?

Teacher's Guide

Look at the pictures with the children and discuss the questions.
See pages 100-1 of the *Teacher's Guide* for key ideas to draw out.

87 ★

Fraction equivalences

Let's learn

$\frac{4}{5}$ add $\frac{3}{10}$ is $\frac{7}{15}$

No, that's not right. If you think about it, $\frac{7}{15}$ is less than $\frac{4}{5}$. If you add $\frac{4}{5}$ and $\frac{3}{10}$ the answer can't be less than $\frac{4}{5}$!

Comparing and ordering fractions

To compare and order fractions, turn them into equivalent fractions with the same denominator. This is called the **common denominator**.

Whatever you do to the denominator, you must do the numerator.

To compare $1\frac{3}{4}$ and $1\frac{4}{5}$, you can ignore the whole numbers as they are the same.

The lowest common denominator for the fractions is 20, because 20 is a multiple of 4 and 5.

$$\frac{3}{4} \times \frac{5}{5} = \frac{15}{20} \qquad \frac{4}{5} \times \frac{4}{4} = \frac{16}{20}$$

So $1\frac{4}{5}$ is larger than $1\frac{3}{4}$.

Adding and subtracting fractions

To add or subtract fractions, turn them into equivalent fractions with a common denominator.
These diagrams show another way of finding common denominators.

$$\frac{1}{2} + \frac{1}{4} = \frac{3}{4}$$

$\frac{1}{2} = \frac{2}{4}$

So

$\frac{2}{4} + \frac{1}{4} = \frac{3}{4}$

$$\frac{2}{3} + \frac{5}{6}$$

$\frac{1}{3} = \frac{2}{6}$

So

$\frac{4}{6} + \frac{5}{6} = \frac{9}{6} = 1\frac{3}{6}$

Teacher's Guide

Before working through the *Textbook*, study page 102 of the *Teacher's Guide* to see how the concepts should be introduced. Read and discuss the page with the children. Provide concrete resources to support exploration.

★**88**

1

Answer these.

Which is the smaller fraction?

a $\frac{3}{4}$ or $\frac{5}{8}$ b $\frac{2}{3}$ or $\frac{5}{6}$ c $1\frac{1}{2}$ or $1\frac{2}{3}$ d $2\frac{2}{3}$ or $2\frac{5}{9}$

Which is the larger fraction?

e $\frac{2}{3}$ or $\frac{4}{5}$ f $\frac{3}{4}$ or $\frac{2}{3}$ g $2\frac{1}{3}$ or $2\frac{3}{5}$ h $3\frac{1}{2}$ or $3\frac{3}{9}$

Remember to find common denominators!

2

Calculate.

Add and subtract fractions.

a $\frac{1}{2} + \frac{5}{6}$ c $\frac{2}{5} + \frac{2}{3}$ e $\frac{7}{8} - \frac{1}{4}$ g $\frac{7}{10} - \frac{1}{4}$

b $\frac{3}{4} + \frac{5}{8}$ d $\frac{5}{6} + \frac{7}{9}$ f $\frac{5}{6} - \frac{2}{3}$ h $\frac{7}{8} - \frac{1}{6}$

3

Apply.

Find the sum of these amounts of money.

a $\frac{2}{5}$ of £1 + $\frac{1}{4}$ of £1

b $2\frac{2}{5}$ of £1 + $1\frac{3}{10}$ of £1

c $2\frac{1}{2}$ of £1 + $2\frac{9}{10}$ of £1

d $3\frac{3}{5}$ of £1 + $3\frac{3}{4}$ of £1

Make the amounts using the fewest coins possible.

Subtract these liquid volumes.

e $5\frac{7}{8}l - 2l$ g $9\frac{5}{6}l - 6\frac{3}{8}l$

f $4\frac{2}{3}l - 3\frac{1}{5}l$ h $10\frac{2}{3}l - 8\frac{1}{4}l$

Round your differences to the nearest 500 ml.
Measure them into containers.

4

Think.

Ali measured 3 jugs of juice.

A $2\frac{3}{4}$ litres

B $2\frac{2}{3}$ litres

C $2\frac{5}{8}$ litres

Which jug has the greatest volume?
Which jug has the least volume?

Teacher's Guide See page 103 of the *Teacher's Guide* for ideas of how to guide practice.
Work through each step together as a class to develop children's
conceptual understanding.

89 ★

Fraction, decimal and percentage equivalences

Let's learn

I think that $\frac{1}{4}$ is the same as 0.4 which is the same as four per cent.

You need:
- coins and notes
- calculators
- paper strips

I'm afraid you are wrong! 0.4 is equivalent to four tenths. $\frac{1}{4}$ is equivalent to $\frac{25}{100}$. We can find that out if we divide 1 by 4. This means that $\frac{1}{4}$ is equivalent to 0.25 which is 25%.

Fractions and division

You can think of fractions as division calculations.

This can help you convert between fractions, decimals and percentages.

Look at these diagrams:

1			
$\frac{1}{2}$		$\frac{1}{2}$	
$\frac{1}{4}$	$\frac{1}{4}$	$\frac{1}{4}$	$\frac{1}{4}$

1			
0.5		0.5	
0.25	0.25	0.25	0.25

100%			
50%		50%	
25%	25%	25%	25%

$$\frac{1}{4}$$

$1 \div 2$ is $\frac{1}{2}$, or 0.5

$1 \div 4$ is $\frac{1}{4}$, or 0.25

$100\% \div 2$ is 50%.

$100\% \div 4$ is 25%.

Finding equivalences mentally

Half of $\frac{1}{2}$ is $\frac{1}{4}$.

So you can find the decimal equivalent of $\frac{1}{4}$ by halving 0.5.

What is half of $\frac{1}{4}$?

Write the fraction, decimal and percentage.

	Fraction	Decimal	Percentage	
halved	$\frac{1}{2}$	0.5	50	halved
halved	$\frac{1}{4}$	0.25	25	halved

$\frac{1}{10}$ is equivalent to 0.1.

Use this fact to work out all the other tenths as decimals.

1 whole is equivalent to 100%.

$\frac{10}{100}$ is equivalent to 10%.

$\frac{1}{100}$ is equivalent to 1%.

Use these facts to write down the equivalent fraction and decimal for 75%.

Teacher's Guide

Before working through the *Textbook*, study page 104 of the *Teacher's Guide* to see how the concepts should be introduced. Read and discuss the page with the children. Provide concrete resources to support exploration.

1

Answer these.

Write these fractions as decimals and percentages.

a $\frac{1}{4}$ c $\frac{1}{2}$ e $\frac{3}{10}$ g $\frac{4}{5}$

b $\frac{1}{10}$ d $\frac{1}{5}$ f $\frac{3}{4}$ h $\frac{3}{8}$

2

Calculate.

If 1 whole is 12, what is each of these?

a 20%   c $\frac{3}{4}$ e $\frac{1}{5}$ g $1\frac{1}{4}$

b 0.5   d 25%   f 1.5   h 2%

3

Solve.

If 100% represents £36, write down what these percentages represent:

a 10%   e 35%

b 5%   f 70%

c 20%   g $17\frac{1}{2}$ %

d $2\frac{1}{2}$ %   h 61%

Now make the amounts using notes and coins.

What is the total of the amount of money you have made?

Use the fewest notes and coins you can.

4

Think.

100% represents £380.

What is 10%?

Write down 5 percentages you can work out from knowing 10%.

Use what you have to make another 5 percentages.

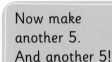

Now make another 5. And another 5!

See page 105 of the *Teacher's Guide* for ideas of how to guide practice. Work through each step together as a class to develop children's conceptual understanding.

91 ★

Formulae

Let's learn

I can find the perimeter of a rectangle by adding all the sides together.

Did you know there is a quicker way? You can use the formula $2l + 2w$ or $2(l + w)$.

Formulae for perimeter

Simple formulae can help you work out problems.

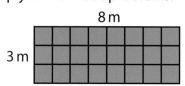

8 m

3 m

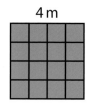

4 m

Perimeter = $2l + 2w$

Or $2(l + w)$

To find the perimeter of the rectangle above you can count the squares or add the sides.

$3\,m + 8\,m + 8\,m + 3\,m = 22\,m$

Or you can add 3 m and 8 m and double.

You can write this as a formula: $P = 2(l + w)$.

A third way is to double 3 and double 8, then add them.

This formula is $P = 2l + 2w$.

Use the formula to find the perimeter of the square.

Can you think of another formula you could use for a square?

Formulae for area

To find the area of the rectangle you can count the squares or add 8 m three times.

Or you can use the formula $A = l \times w$.

Use the formula to find the area of the square.

6 m

Area = $l \times w$

How could you use the formula of the 6 metre square to work out the formula of the triangle and parallelogram?

Teacher's Guide

Before working through the *Textbook*, study page 106 of the *Teacher's Guide* to see how the concepts should be introduced. Read and discuss the page with the children. Provide concrete resources to support exploration.

★92

1

Draw.

Draw squares with these side lengths.
Work out their areas using the formula.

a   4 cm

c   12 cm

e   20 cm

g   90 mm

b   6 cm

d   15 cm

f   72 mm

h   115 mm

2

Draw.

Draw rectangles with these dimensions. Work out their perimeters and areas using the formulae.

a   6 cm by 4 cm

d   15 cm by 5 cm

g   9.6 cm × 15 cm

b   10 cm by 3 cm

e   10.5 cm by 12 cm

h   10.5 cm × 20 cm

c   12 cm by 2 cm

f   8.5 cm by 15 cm

3

Apply.

Mr Green designed a patio for his garden. It looked like this:

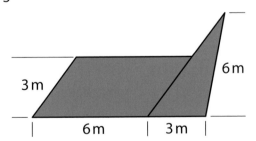

Work out the area of the patio so that Mr Green can order the correct amount of paving.

Design a patio for Mr Green using 2 parallelograms and 2 triangles. Measure each shape in centimetres and scale these up to metres. What is the area of your patio?

4

Think.

Eva drew a rectangle.

It had a perimeter of 24 cm.

Draw all the rectangles that Eva could have drawn.

Now work out their areas.

Teacher's Guide

See page 107 of the *Teacher's Guide* for ideas of how to guide practice. Work through each step together as a class to develop children's conceptual understanding.

93 ★

Missing number statements

Let's learn

I think the missing number is 14.

5 + 9 = ▪ + 6

It can't be! The numbers on both sides of the equals symbol must be equivalent. In your calculation 5 add 9 would be equivalent to 14 add 6, which is 20! This isn't right.

Find missing numbers using the bar model

Look at the bar model. It shows that $a + b$ is equivalent to $c + d$.

Use the model to find c in this statement:
$12 + 4 = c + 10$

Find a in this statement: $a + 7 = 9 + 12$

Now find b: $25 + b = 17 + 19$

Find d: $23 + 13 = 15 + d$

Make up some statements of your own.

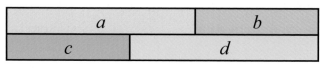

$$a + b = c + d$$

Find missing numbers using balancing

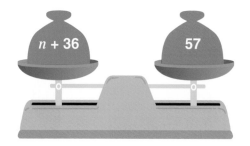

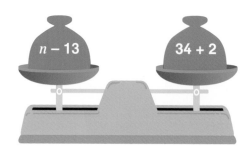

Look at this number statement (equation): $n + 36 = 57$

What is n?

One way to find out is to balance the statement (or equation).

Remember, the equals symbol shows equivalence. What you do to one side must be done to the other.

Take away 36 from both sides of the equals symbol and you have $n = 21$.

Look at the second balance. What is n?

Teacher's Guide

Before working through the *Textbook*, study page 108 of the *Teacher's Guide* to see how the concepts should be introduced. Read and discuss the page with the children. Provide concrete resources to support exploration.

1

Answer these.

Find the value of the unknown numbers by drawing the bar models.

a $m + 35 = 72$

b $54 = 27 + n$

c $p - 46 = 27$

d $74 + s = 145$

e $m + 250 = 362$

f $p + 263 = 421 + 124$

g $254 + s = 416 + 124$

h $138 - n = 87$

2

Answer these.

Find the value of the unknown numbers by balancing.

a $23 + n = 48$

b $14 + n = 27$

c $n + 36 = 43 - 4$

d $2n + 15 = 35$

e $48 = 2n + 12$

f $35 + 2n - 21$

g $n + 30 = 3n$

h $4n + 14 = 2n - 14$

You can use words or pictures.

3

Measure.

Measure and cut 2 cm wide strips of paper to these lengths.

a   6 cm

b   4.5 cm

c   23 mm

d $12\frac{1}{5}$ cm

Use pairs of strips to create some missing number statements, e.g.:

☐ + strip of 23 mm = 6 cm.

Each time measure and cut a new strip to show the missing number.

Check your answer.

4

Think.

Copy this diagram.

Place each of the numbers 1 to 5 in one of the circles.

Each diagonal of circles must have the same total.

How many possible ways can you do this?

What do you notice?

How can you be sure you have all the possibilities?

Teacher's Guide

See page 109 of the *Teacher's Guide* for ideas of how to guide practice.
Work through each step together as a class to develop children's conceptual understanding.

95 ★

Unknown numbers

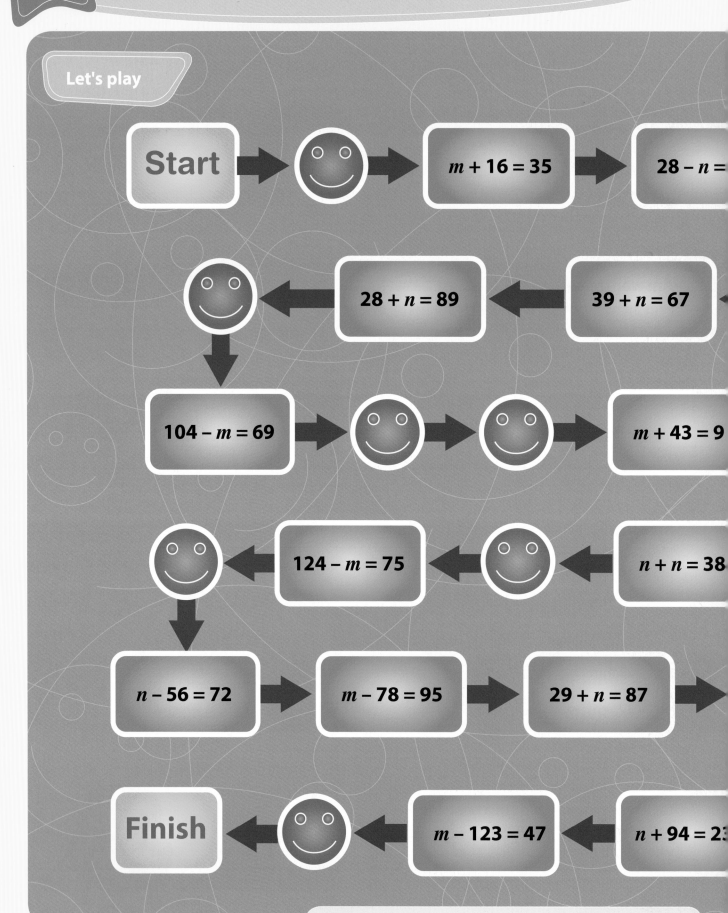

Let's play

Start

$m + 16 = 35$

$28 - n =$

$28 + n = 89$

$39 + n = 67$

$104 - m = 69$

$m + 43 = 9$

$124 - m = 75$

$n + n = 38$

$n - 56 = 72$

$m - 78 = 95$

$29 + n = 87$

Finish

$m - 123 = 47$

$n + 94 = 23$

Teacher's Guide

See pages 110-11 of the *Teacher's Guide*. Explain the rules for each game and allow children to choose which to play. Encourage them to challenge themselves and practise what they have learnt in the unit.

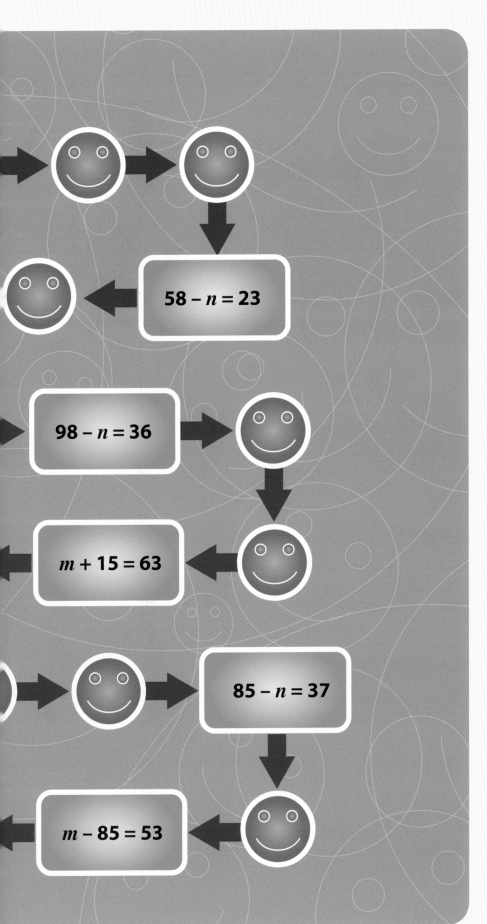

58 − n = 23

98 − n = 36

m + 15 = 63

85 − n = 37

m − 85 = 53

1 What is it?

Find the unknowns and make the highest total to win the game.

2 How to make it right

Rewrite the number sentences to make them correct and score points.

3 Your game

Design your own game using the gameboard. Explain the rules and play with a partner.

And finally ...

1

Which is largest?

a $2\frac{3}{5}$ or $2\frac{4}{6}$

b $5\frac{7}{10}$ or $5\frac{4}{5}$

c $\frac{22}{3}$ or $\frac{29}{4}$

d $\frac{40}{6}$ or $\frac{34}{5}$

Make sure you have changed any improper fraction answers to mixed numbers.

Add or subtract.

e $\frac{1}{2} + \frac{3}{4}$

f $\frac{4}{5} + \frac{7}{10}$

g $\frac{7}{8} - \frac{1}{4}$

h $\frac{7}{12} - \frac{1}{3}$

i $\frac{3}{5} + \frac{5}{6}$

j $\frac{5}{6} - \frac{3}{8}$

k $\frac{2}{3} + \frac{9}{12}$

l $\frac{7}{9} - \frac{2}{3}$

Convert to decimals and percentages.

m $\frac{3}{5}$ n $\frac{7}{10}$ o $\frac{3}{4}$ p $\frac{3}{8}$

Convert to fractions and percentages.

q   0.2   r   0.75   s   0.06   t   1.15

Remember to simplify any fractions that can be simplified.

Convert to fractions and decimals.

u   30%   v   95%   w   48%   x $12\frac{1}{2}$%

Teacher's Guide

See pages 112–13 of the *Teacher's Guide* for guidance on running each task. Observe children to identify those who have mastered concepts and those who require further consolidation.

2 Find the unknown numbers.

a $m + 23 = 50$

b $y - 13 = 28$

c $25 + n = 72$

d $36 - s = 12$

e $s + 36 = 98$

f $t - 47 = 28$

g $m - 56 = 40$

h $n + 27 = 95$

i $56 - v = 24.5$

Draw a triangle and a parallelogram.
Find their perimeters and areas.
Write the formula for the area inside each shape.

Draw another 2 and work out their perimeters and areas.

Now draw another 2.

You need:
- squared paper
- ruler

What do you notice about the perimeters and areas of the triangles and parallelograms you drew?

Did you know?

Algebra has been around for centuries. In early civilisations people used to visualise things they were not sure of or problems that they wanted to solve.

That's right. Over time these visualisations became actual representations and abbreviations. Finally, in the 17th century they became algebraic symbols.

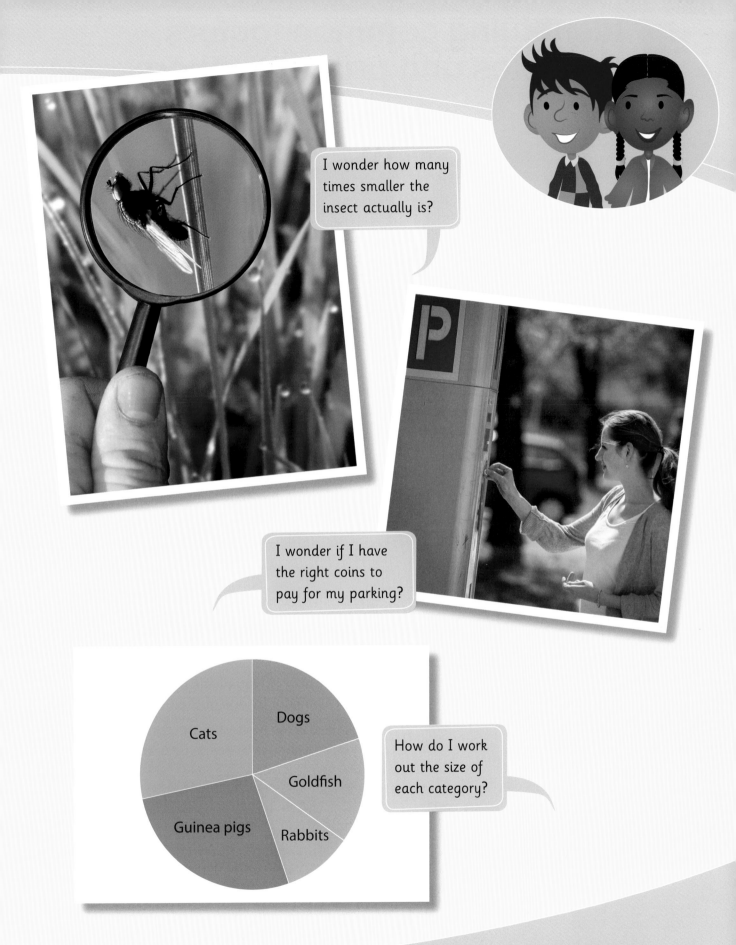

I wonder how many times smaller the insect actually is?

I wonder if I have the right coins to pay for my parking?

How do I work out the size of each category?

Cats

Dogs

Goldfish

Guinea pigs

Rabbits

Teacher's Guide   Look at the pictures with the children and discuss the questions.
See pages 114-15 of the *Teacher's Guide* for key ideas to draw out.

101 ★

Identifying common factors, multiples and prime numbers

Let's learn

I simplified $\frac{36}{60}$ by halving the numerator and denominator to get $\frac{18}{30}$. I halved them again to get $\frac{9}{15}$. I can't halve anymore, so the fraction is in its simplest form.

That's not in its simplest form yet. The numerator and denominator still have a common factor of 3. Divide them both by 3 to get $\frac{3}{5}$. The fraction is now in its simplest form.

You need:
- adhesive notes
- calculator
- multiplication table squares
- counters

Finding factors

Look at 48×13.

13 is prime, so it has no proper factors.

48 is not prime, so it can be written as the product of a factor pair.

$6 \times 8 \times 13$ is one way to rewrite the product.

$2 \times 2 \times 2 \times 2 \times 3 \times 13$ is another way to rewrite the product.

Common factors

Look at 18 and 24.

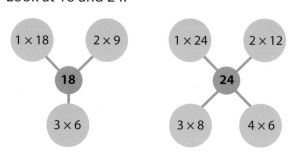

The **factors** of 18 are 1, 2, 3, 6, 9 and 18.

The factors of 24 are 1, 2, 3, 4, 6, 8, 12 and 24.

To simplify $\frac{18}{24}$, you divide the numerator and denominator by one of their common factors.

Since 6 is the highest of the common factors, you can choose to divide by 6, so $\frac{18^{3}}{24_{4}} = \frac{3}{4}$.

$\frac{3}{4}$ is the simplest form of $\frac{18}{24}$.

Common multiples

The **multiples** of 18 are 18, 36, 54, 72 …

The multiples of 24 are 24, 48, 72, 96 …

72 is a multiple of 18 and 24.

To compare $\frac{11}{18}$ and $\frac{13}{24}$, you can write them as fractions with the same denominator.

You can use the lowest common multiple, which is 72.

$\frac{11}{18} = \frac{11}{18} \times \frac{4}{4} = \frac{44}{72}$ and $\frac{13}{24} = \frac{13}{24} \times \frac{3}{3} = \frac{39}{72}$.

So, $\frac{11}{18}$ is larger than $\frac{13}{24}$.

Teacher's Guide

Before working through the *Textbook*, study page 116 of the *Teacher's Guide* to see how the concepts should be introduced. Read and discuss the page with the children. Provide concrete resources to support exploration.

1

Answer these.

a   Write 6 as the product of prime numbers.

b   Write 84 as the product of prime numbers.

c   Find all the common factors of 16 and 40.

d   Find all the common factors of 12, 24 and 45.

e   Find 2 common multiples of 8 and 10.

f   Find 2 common multiples of 15, 16 and 24.

2

Answer these.

Work out the answers using primes, common factors or common multiples.

a 42×31

b 115×14

c   Simplify $\frac{35}{42}$

d   Simplify $\frac{48}{72}$

e   Order $\frac{5}{6}$ and $\frac{7}{10}$, with the smaller one first.

f   Order $\frac{2}{3}$, $\frac{5}{8}$ and $\frac{3}{7}$, with the smallest one first.

3

Apply.

a   Pencils are sold in packs of 10. Rulers are sold in packs of 6. How many packs of each do I need in order to have the same number of pencils and rulers?

b   There are 30 children in my class. I have 12 sets of mathematical equipment. I want to arrange the tables in groups so that the equipment is shared equally among children. How many groups do I need? The groups should be equal.

c   There are 32 children in my class. Fun-size treats come in bags of 20. What is the smallest number of bags I must buy so that each child has the same number of treats?

d   A swimming pool is 25 m long. I swim 28 lengths. How far do I swim?

e   A box measures 24 cm by 15 cm by 25 cm high. Work out its volume.

4

Think.

I am thinking of 2 different numbers. They have 6 as a common factor and 180 as a common multiple. What are the numbers?

a   Make up 3 more problems like this. Is there always just one answer? Is there a quick way to find answers to questions like these?

b   Which 2-digit number has the most factors?

Teacher's Guide

See page 117 of the *Teacher's Guide* for ideas of how to guide practice. Work through each step together as a class to develop children's conceptual understanding.

103 ★

Multiplying and dividing decimal numbers

Let's learn

You need:
- calculators
- Gattegno charts
- place-value counters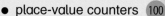

I worked out 3.076 × 1000. It is 3.076000 because you add 3 zeros to multiply by 1000.

Your answer is the same as the number you started with! Multiplying by 1000 should make it much bigger. 3 ones should become 3 thousands. 7 hundredths becomes 7 tens and 6 thousandths becomes 6 ones. This gives 3076.

Multiplying and dividing by 10, 100 and 1000

Look at the Gattegno chart.

0.001	0.002	0.003	0.004	0.005	0.006	0.007	0.008	0.009
0.01	0.02	0.03	0.04	0.05	0.06	0.07	0.08	0.09
0.1	0.2	0.3	0.4	0.5	0.6	0.7	0.8	0.9
1	2	3	4	5	6	7	8	9
10	20	30	40	50	60	70	80	90
100	200	300	400	500	600	700	800	900
1000	2000	3000	4000	5000	6000	7000	8000	9000

Each row of numbers is 10 times larger than the row above. When multiplying, move down the chart.

You need to move 1 row down to multiply by 10 and 2 rows down to multiply by 100.

How would you use the chart to multiply by 1000?

For division, you need to move up the chart. Move up once to divide by 10, twice to divide by 100 and 3 times to divide by 1000.

Multiplying decimals with answers up to three decimal places

Look at 3.71 × 8.

Use short multiplication:

This works in the same way as with whole numbers.

$$\begin{array}{r} 3.71 \\ \times 8 \\ \hline 29.68 \\ _2\,_5 \end{array}$$

Check the position of the decimal point by estimating.

4 × 8 = 32, so the answer should be a little less than 32.

Dividing decimals with answers up to three decimal places

Look at 3.934 ÷ 7.

Use short division:

This works in the same way as with whole numbers.

$$\begin{array}{r} 0.562 \\ 7\overline{)3.^3 9^4 3\,^1 4} \end{array}$$

Check by estimating using inverse operations.

7 × 0.6 = 4.2, so the starting number to be divided must have been a bit less than 4.2.

Teacher's Guide

Before working through the *Textbook*, study page 118 of the *Teacher's Guide* to see how the concepts should be introduced. Read and discuss the page with the children. Provide concrete resources to support exploration.

1

Solve.

a 6.732×100 c $48\,091 \div 1000$ e 3.784×6 g $6.78 \div 12$

b $256 \div 100$ d 0.038×1000 f $5.327 \div 7$ h $271 \div 8$

2

Copy and complete.

a   ▢ $\times 100 = 496$ c $5307 = 5.307$ ▢ 1000 e $2.25 \times$ ▢ $= 13.5$

b $2.301 =$ ▢ $\div 100$ d $4 \div$ ▢ $= 0.04$ f   ▢ $\div 6 = 1.09$

3

Apply.

The graph shows the height of a broad bean plant at the end of each week, from when it is planted.

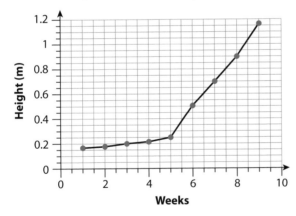

a   What is its height in centimetres at the end of the third week?

b   How many centimetres did the broad bean plant grow in the eighth week from planting?

c   How much, in metres, did the plant grow between week 3 and week 7?

d   When did the plant first start growing much more rapidly?

4

Think.

I am thinking of two 2-digit numbers. They divide to give 3.625. What are the two numbers?

Is there more than one answer?

How do you know?

Teacher's Guide

See page 119 of the *Teacher's Guide* for ideas of how to guide practice. Work through each step together as a class to develop children's conceptual understanding.

105 ⭐

Solving problems with percentages

You need:
- 100 squares

I can buy a new pair of shoes for £5. The full price is £15 and there is a sale on. There is 10% off in the sale, so the shoes will only cost £5.

You're hopeful! 10% off means 10% of £15. This is £1.50, not £10! Your shoes will cost £13.50.

Calculating percentages of amounts

To calculate 24% of 120, use a bar to represent 100%.

120 is the whole, so it matches 100%.

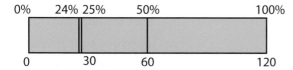

50% is 60, so 25% is 30.

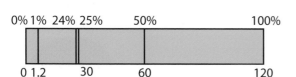

25% is 1% away from 24%. 1% of 120 is 1.2.

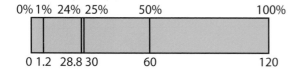

This gives 24% of 120 = 30 − 1.2
$$= 28.8$$

Working out the percentage

To work out what percentage 18 is of 30, use a bar to represent 100%.

30 is the whole, so it matches 100%.

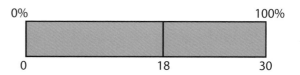

Deduce that 6 is 20% by dividing both by 5.

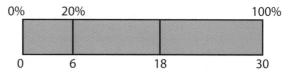

Deduce that 18 is 60% by multiplying 20% and 6 by 3.

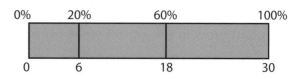

So 18 is 60% of 30.

Teacher's Guide

Before working through the *Textbook*, study page 120 of the *Teacher's Guide* to see how the concepts should be introduced. Read and discuss the page with the children. Provide concrete resources to support exploration.

1 Solve.

a   12% of 300

b   20% of 150

c   5% of 240

d   15% of 360

e   46% of 600

f   55% of 130

g   What percentage is 8 of 20?

h   What percentage of 40 is 24?

i   What is 85 out of 500 as a percentage?

j   Work out 19 out of 76 as a percentage.

2 Copy and complete.

a   20% of ▢ is 17

b   8 is 40% of ▢

c   6 is $2\frac{1}{2}$% of ▢

d   32 is ▢ % of 40

e   11 is ▢ % of 20

f   60% of ▢ is 45

3 Apply.

Favourite colour	Percentage
Red	15
Yellow	35
Blue	30
Green	20

a   Ali draws a pie chart of this information:
Work out the angle in the pie chart for blue.

Work out the angle in the pie chart for yellow.

b   Eva has £6 pocket money. She saves 35% of it. How much does she save?

c   Ali spends 6 hours at school. He is sitting for 65% of the time. How long is that in hours and minutes?

d   A holiday costs £450. There is a discount of 10% if you book online. How much is the saving for booking online?

e   Ali walks 800 m to his friend's house. It starts to rain when he has travelled 80% of the way. How much further has he to go?

f   Eva spends £1.80 of her £6 pocket money. What percentage is that?

4 Think.

Ali sees an offer on a computer game. There is 20% off the price. He also has a voucher for 10% off. The game costs £15. He thinks he will save £4.50, but he is wrong. How much does he save? Is it better to use the voucher before the shop's reduction or afterwards?

Now make up a similar problem of your own for others to solve.

Teacher's Guide

See page 121 of the *Teacher's Guide* for ideas of how to guide practice. Work through each step together as a class to develop children's conceptual understanding.

107 ★

Solving equations

You need:
- number rods
- digit cards

Let's learn

I'm solving equations. This one says that $A + B = 7$. That means $A = 3$ and $B = 4$ because B is one after A in the alphabet.

That is one answer, your reasoning isn't right. The alphabet order doesn't mean anything in mathematics. Here A and B can stand for any numbers that add to give 7. You could have $A = 5$ and $B = 2$.

Satisfying equations

Look at $a + b = 6$.

a and b are positive whole numbers.

What could a and b stand for?

6	
a	b

$a = 4$ and $b = 2$ is one solution.

6	
4	2

So are all of these.

6	
1	5
2	4
3	3
4	2
5	1

Look at $a + a + a + b = 11$.

a and b are positive whole numbers.

11			
a	a	a	b

What could a and b stand for?

What are the possibilities?

Look at Ali's equation.
A can take the values 1, 2, 3 and 4.
B can take the values 1, 2 and 3.

B \ A	1	2	3	4
1	1,1	1,2	1,3	1,4
2	2,1	2,2	2,3	2,4
3	3,1	3,2	3,3	3,4

The table shows all the possible pairs of numbers for A and B.

How many ways can you choose 2 different numbers?

This shows all the pairs, but some are the same, e.g. 1 and 3 are the same pair as 3 and 1.

The table leaves 6 possible pairs of numbers.

1st \ 2nd	1	2	3	4
1	1,1	1,2	1,3	1,4
2	2,1	2,2	2,3	2,4
3	3,1	3,2	3,3	3,4
4	4,1	4,2	4,3	4,4

Teacher's Guide

Before working through the *Textbook*, study page 122 of the *Teacher's Guide* to see how the concepts should be introduced. Read and discuss the page with the children. Provide concrete resources to support exploration.

★108

1

Answer these.

Identify the unknowns.

a p and q are positive whole numbers.
$p + q = 3$
List the pairs of values that p and q could have.

b p and q are positive whole numbers.
$p + p + q = 10$
List the pairs of values could p and q could have.

c   Choose 2 digit cards. List the different ways you can do this. The order of the numbers does **not** matter.

d   Choose 2 digit cards to make a 2-digit number.
List the different ways you can do this. The order of the numbers **does** matter.

2

Solve.

a $p = 4$ and $q = 5$ and so $p + q = 9$.
List 12 pairs of values for p and q that also have a sum of 9.

b $p - q = 7$ and $p + q < 14$
p and q are both positive whole numbers. List the possible pairs of values for p and q.

c   Choose some digit cards. Work out their sum. List the numbers you can make. You may use just 1 number. You may not use any number more than once in a sum.

3

Solve.

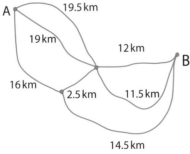

a   What is the shortest route from A to B? Justify your answer.

b   The distance marked 2.5 km is now 3.5 km. What is the shortest route now?

c   The distance marked 12 km is now 15 km. What is the shortest route now?

d   The distance marked 19 km is now 18 km. What is the shortest route now?

4

Think.

a   You only have 2 values of stamps, 3p and 8p. What is the largest value that cannot be made using as many as you like of these stamps?

b   The stamps are now worth 5p and 6p. What is the largest value that cannot be made?

Teacher's Guide
See page 123 of the *Teacher's Guide* for ideas of how to guide practice. Work through each step together as a class to develop children's conceptual understanding.

109 ★

Challenging numbers

Let's play

1 Start

2

3 move forward **2** spaces

4

5 doubl you scor

12 move forward **2** spaces

11

10 doub you scor

13

14 go back **3** spaces

15 double your score

16

17

Mission ①

Find a common factor of your number and the challenge number.

Mission ②

Find a common multiple of your number and the challenge number.

Mission ③

Your number is a percentage. Work out that percentage of the challenge number.

Teacher's Guide
See pages 124-5 of the *Teacher's Guide*. Explain the rules for each game and allow children to choose which to play. Encourage them to challenge themselves and practise what they have learnt in the unit.

★110

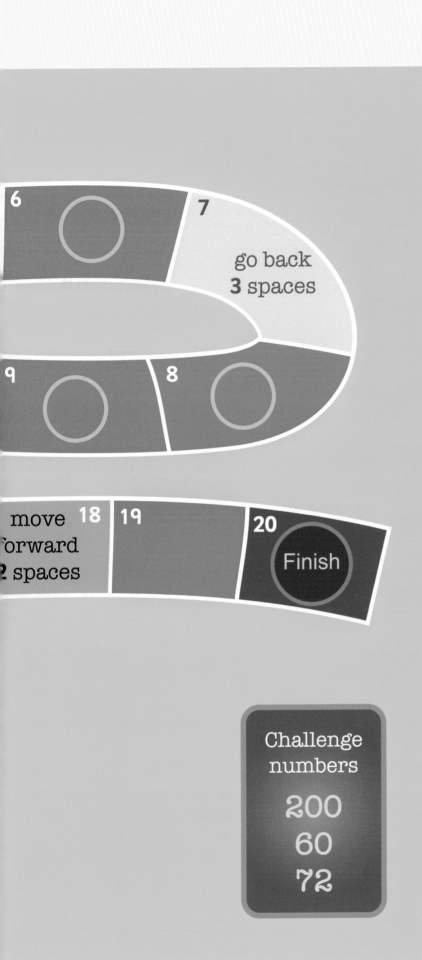

6

7 go back **3** spaces

9

8

move 18 **19**
forward
2 spaces

20 Finish

Challenge
numbers
200
60
72

You need:

- 1–6 dice
- counters
- calculator

1 Challenging tens

Make a tens number, choose a challenge number and carry out your mission. Who can reach the end of the track first?

2 2-digit challenge

Make a 2-digit number, choose a challenge number and carry out your mission. Who can reach the end of the track first?

3 Your game

Make up your own game using the gameboard. Explain the rules and play with a partner.

Review 8

And finally ...

Let's review

1 Someone has answered these questions and made some mistakes.
Mark their work and write some feedback to help them understand where they went wrong.

a Work out 12% of 60.

12% of 60 = 72

b Work out 8×1.4.

8 x 1.4 = 8.32

c Work out $5.406 \div 3$.

5.406 ÷ 3 = 1.82

d Work out the possible values for a and b if $a + a + b = 7$.

a = 2 and b = 3 is the only solution

e Two numbers have a sum of 12 and a product of 27. What are the two numbers?

4 and 8 make 12 but 4 x 8 = 32, so it can't be done

f Work out the possible values of a and b if $a + b = 5$.

1a + 4b, 2a + 3b, 3a + 2b, 4a + 1b

g Write 15 as a product of prime numbers.

15 = 5 x 3 x 1

Teacher's Guide

See pages 126–7 of the *Teacher's Guide* for guidance on running each task. Observe children to identify those who have mastered concepts and those who require further consolidation.

2

Any number can be written as the product of primes.

28 = 2 × 2 × 7 as a product of prime numbers.

What is 28 as the sum of prime numbers?

Use as few prime numbers in your sum as possible.

Is there more than one way?

What about 102?

I wonder if every number can be written as the sum of prime numbers?

3

The number of children whose birthdays occur in each season in a school is shown in the table:

Season	Frequency
Autumn	108
Winter	24
Spring	60
Summer	48

a   Work out what percentage of children are born in each season. Explain how you arrived at your answers.

b   In another school there are 150 children. 32% of the children are born in the autumn. How many of them are born in the autumn?

c   In a third school 20% are born in the autumn. There are 36 children born in the autumn. How many children are in the school altogether?

Did you know?

1 used to be a prime number. In the nineteenth century, it was included in the list of prime numbers. Later it was removed because it didn't support several important theorems about prime numbers.

Prime numbers are used to encode data on the Internet. It is difficult to decide if a very large number is prime. You basically have to divide by all the prime numbers up to the square root of the number. It can take several years using many computers working together on it.

Can you describe the circles in this old watch mechanism?

How do the sizes compare?

How can you compare the sizes of these boxes?

I wonder how you could describe the position of these people?

Teacher's Guide
Look at the pictures with the children and discuss the questions.
See pages 128-9 of the *Teacher's Guide* for key ideas to draw out.

115 ⭐

Circles and scaling

Let's learn

The area of this triangle is 12 cm². When it is enlarged by a scale factor of 3, the new area will be 3 times greater, making it 36 cm².

Remember the area of a triangle equals $\frac{1}{2}(bh)$. I think the area will be 9 times greater. The base is 3 times bigger and the height is 3 times bigger so the area will be (3 × 3) times bigger.

Enlarging with scale factors

The size of an enlargement is described by its scale factor.

This circle and triangle have been enlarged by a scale factor of 2 because each measurement is twice as long.

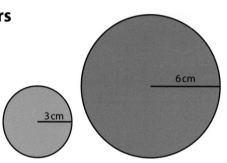

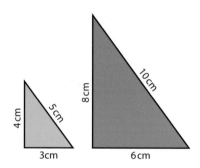

Naming parts of a circle

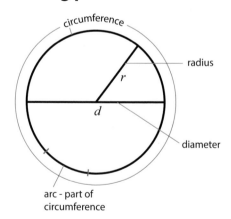

arc - part of circumference

The **circumference** of a circle is the distance around the edge. It can be described as the perimeter of a circle.

An **arc** is any part of the circumference.

The **radius** is the distance from the centre of a circle to the circumference.

The **diameter** is a straight line from edge to edge, passing through the centre.

So, the diameter is twice the radius, $d = 2r$.

Open your compass to the radius of a circle. The distance fits around the circumference exactly 6 times. This is useful for drawing a hexagon or equilateral triangle inside a circle.

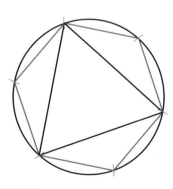

Teacher's Guide

Before working through the *Textbook*, study page 130 of the *Teacher's Guide* to see how the concepts should be introduced. Read and discuss the page with the children. Provide concrete resources to support exploration.

1

Copy and complete.

Copy the table.
Complete the first
2 columns using the
rule $d = 2r$.

Now complete the rest
of the table when the
circles are increased by
a scale factor of 3.

		Increase by a scale factor of 3	
Radius	**Diameter**	**New radius**	**New diameter**
9 cm			
4 m			
	24 cm		
7.5 cm			
	1 m		

2

Calculate.

Calculate the radius of these circles.

a $d = 4$ cm
         b $d = 12$ cm
         c $d = 9$ cm

Use compasses to draw each circle. Measure the diameter to check.

Use a scale factor of 2 to increase the size of the first circle. Draw the enlargement.

3

Apply.

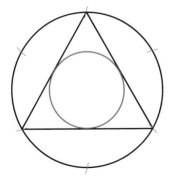

Work in a small group. Use compasses
to draw a simple design similar to this.
Decide on a scale factor and calculate the
new design size.

Create your design in the playground.
Use string and chalk to draw any circles.
Use a metre stick to draw any triangles.

4

Think.

Use angle facts to calculate the
angle sizes in this diagram.

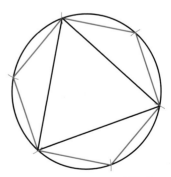

Look at the angles
in the design in
Step 3. Try to
calculate them.

Teacher's Guide

See page 131 of the *Teacher's Guide* for ideas of how to guide practice.
Work through each step together as a class to develop children's
conceptual understanding.

117

Finding missing values

The 3 angles in a triangle add up to 180°, so each angle is 60°.

Not always! That's only in equilateral triangles. For example, a right-angled triangle has one angle that is 90°.

Finding missing angles

To find missing angles use clues:

- geometry symbols – small dashes show lines of equal length. The first set of equal lines has a single dash, the next set has 2 dashes and so on.
- known angle facts – angles in a triangle total 180°.
- known properties of shapes – the angles in a rectangle are 90°.

The diagram shows one way of working out the sizes of all the missing angles. There are other ways to reach the same answers.

This angle is 180° – (90° + 20°) = 70° because the angles in a triangle total 180°.

This angle is 40° because the angles in a triangle total 180° so 180° – (70° + 70°) = 40°.

This angle is 70° because the dashes show that it is an isosceles triangle.

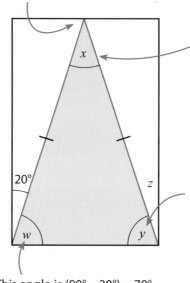

This angle is (90° – 20°) = 70° because the interior angle of a rectangle is 90°.

Finding missing lengths

Find the height of this box.

First, change the width from mm to cm, so that you are working with the same units.

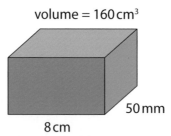

volume = 160 cm³

50 mm

8 cm

Width = 50 mm = 5 cm

Volume, $V, = l \times w \times h$

Substitute the values into the equation:

$160 = 8 \times 5 \times h$

$160 = 40 \times h$

Divide both sides of the equation by 40:

$h = 4$.

The height of the box is 4 cm.

Teacher's Guide

Before working through the *Textbook*, study page 132 of the *Teacher's Guide* to see how the concepts should be introduced. Read and discuss the page with the children. Provide concrete resources to support exploration.

1 **Calculate.**

Copy these diagrams.
Work out the value of each
angle inside the rectangles
using the geometry symbols
and angle facts. Explain
your thinking.

Lines with the same number of
dashes on them are equal in length.

a

b

2 **Calculate.**

3 different cuboids have the same volume, 270 cm³.

a   Find the height.

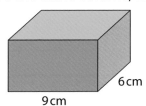

6 cm
9 cm

b   Find the length if the width is
9 cm and the height 3 cm.

c   Find the width if the length is
15 cm and the height 20 mm.

3 **Solve.**

a   Ali's birthday cake is 20 cm long,
10 cm wide and 5 cm deep. He cuts
the cake into 20 pieces. Calculate
the volume of each piece.

b   A can of fizzy drink is 300 ml. Ali has
a pack of 24 cans for his birthday
party. What is the total volume
of the pack of 24 cans? Give your
answer in litres and millilitres.

c   Ali has 3 presents for his birthday.
Each present has a total volume of
480 cm³. Calculate possible values
of length, width and height for the
presents.

4

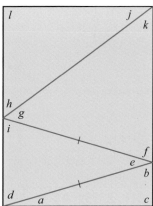

l   j
k
h
g
i
f
e
b
d
a   c

Think.

Without using a
protractor suggest
a possible solution
for the angles in
the diagram

Which angle(s)
would you need
to know in order
to calculate exact
answers?

Discuss your findings
with a friend.

Teacher's Guide

See page 133 of the *Teacher's Guide* for ideas of how to guide practice.
Work through each step together as a class to develop children's
conceptual understanding.

Translation over four quadrants

Let's learn

In a full coordinate grid, the second quadrant appears below the first quadrant.

No, you're wrong! The quadrants are numbered in an anti-clockwise direction. The ancient Babylonians decided the direction based on astronomy. You just have to remember it!

Using four quadrants

The point (0, 0) is called the origin.

Values on the x-axis to the left of the origin are negative. This is the same as on a number line.

On the y-axis, values above the origin are positive and below are negative.

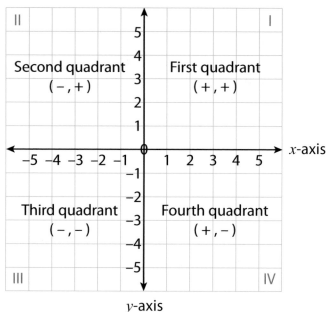

Translating using a full coordinate grid

The kite in the second quadrant has been translated using + 4 in an x direction and – 3 in a y direction.

(–3, 3) (1, 0)

(–4, 2) (0, –1)

(–3, 0) (1, –3)

(–2, 2) (2, –1)

Each point has moved 4 squares to the right (in a positive direction) and 3 squares down (in a negative direction).

The 2 kites are **congruent** (or identical).

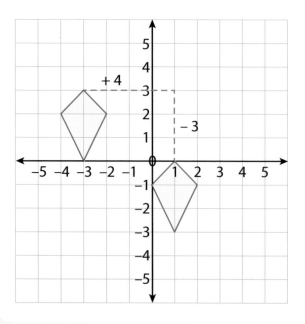

Teacher's Guide

Before working through the *Textbook*, study page 134 of the *Teacher's Guide* to see how the concepts should be introduced. Read and discuss the page with the children. Provide concrete resources to support exploration.

1

Answer.

Describe how these shapes have been translated using x and y.

a

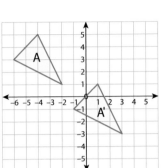

b

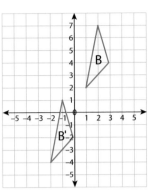

c

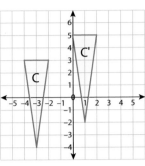

2

Answer.

Use the coordinates marked on the quadrants to make a kite, square, trapezium and rectangle.

Write down their coordinates. (Some coordinates are used more than once.)

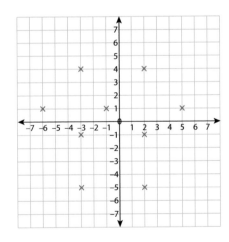

3

Draw.

Draw the outline of a simple robot on a coordinate grid. Use all 4 quadrants (+10 to −10).

Write a list of the coordinates in the order they should be joined.

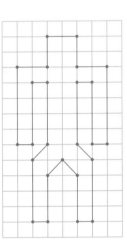

Challenge your partner to plot the coordinates and draw your robot.

4

Think.

Use the letters at the coordinates to crack this message.

Write your own message of 5 or more words.

You will need to add more coordinates for other letters.

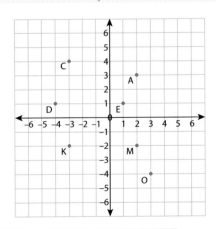

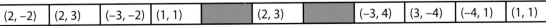

| (2, −2) | (2, 3) | (−3, −2) | (1, 1) | | (2, 3) | | (−3, 4) | (3, −4) | (−4, 1) | (1, 1) |

Teacher's Guide

See page 135 of the *Teacher's Guide* for ideas of how to guide practice. Work through each step together as a class to develop children's conceptual understanding.

121 ★

Get coordinated!

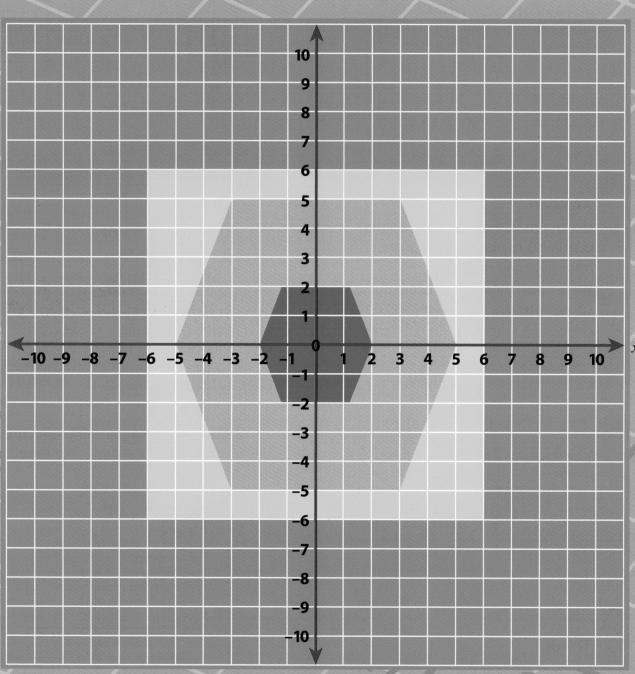

Teacher's Guide

See pages 136-7 of the *Teacher's Guide*. Explain the rules for each game and allow children to choose which to play. Encourage them to challenge themselves and practise what they have learnt in the unit.

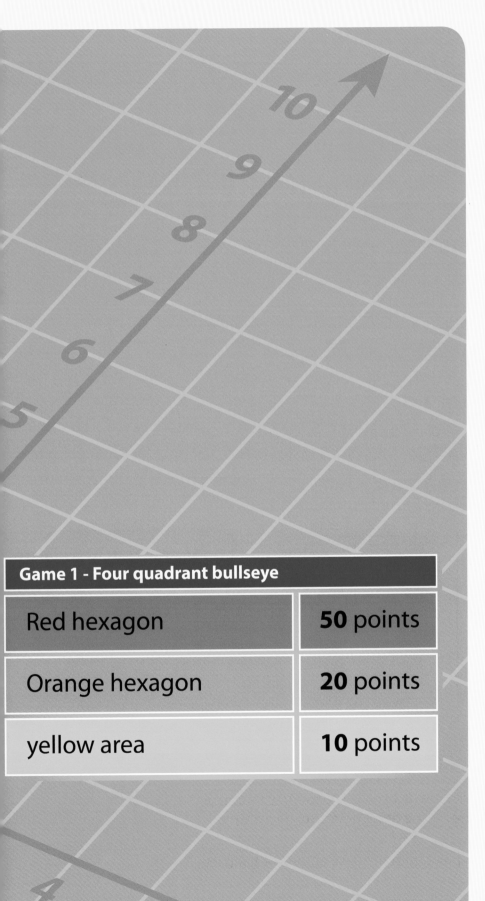

Game 1 - Four quadrant bullseye	
Red hexagon	**50** points
Orange hexagon	**20** points
yellow area	**10** points

You need:

- 1–6 dice (1 red and 1 blue)
- 2 +/– dice (or 2 coins)
- counters

1 Four quadrant bullseye

Play to win by finding and scoring coordinates across the 4 quadrants.

 Dice roll translations

Carry out translations using dice throws. Who will score the most points?

 Your game

Design your own game using the gameboard. Explain the rules and play with a partner.

Let's review

1

Copy and use the geoboard to make different angles. Calculate the size of the angles.

Explain how you worked out each angle.

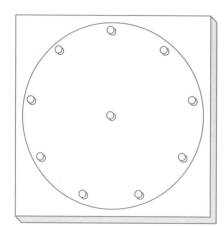

You need:
- compasses
- protractor
- ruler

2

To draw this capital letter M you need to join 5 points.

If the middle point has the coordinates (0, 0) on a +10 to −10 coordinate grid, estimate possible coordinates of the other points to draw a large version of this letter.

Work out the translation of each point from the base of the left-hand leg.

You need:
- squared paper

Teacher's Guide

See pages 138-9 of the *Teacher's Guide* for guidance on running each task. Observe children to identify those who have mastered concepts and those who require further consolidation.

Eva and Ali answered a question about scale factors. They have made some mistakes.

Work out the correct answers, then explain their errors.

a   A right-angled triangle has sides 1.5 cm, 2.0 cm and 2.5 cm. An enlarged
right-angled triangle has sides 6 cm, 7.5 cm and 4.5 cm. What is the scale factor?

The scale factor is 4.

b   A rectangle 3 cm × 1 cm is enlarged by a scale factor of 4. How many times
greater is the area of the new rectangle?

The area is 4 times greater.

c   Enlarge a circle with a radius 3 cm by a scale factor of 4.
What is the diameter of the new circle?

The diameter is 12 cm.

Did you know?

Crop circles are large, intricate circular patterns created by flattening crops. They often appear overnight. At first people believed that aliens created them!

They're probably man-made by pranksters. In 1991, 2 men admitted that they were responsible for many crop circles in England. They even made one in front of journalists, using a plank of wood to flatten the crops and a rope as a pair of compasses. The circles are still a bit mysterious though.

Focus on algebra

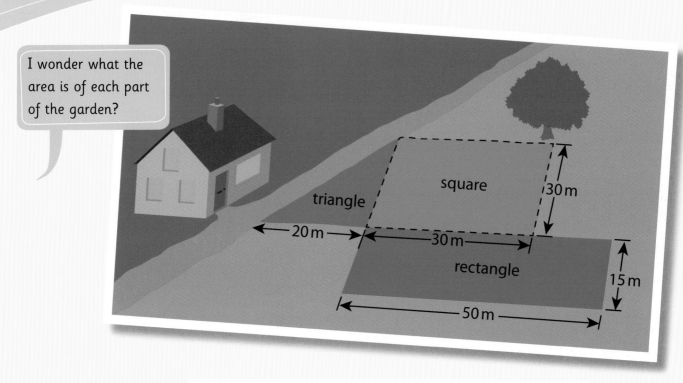

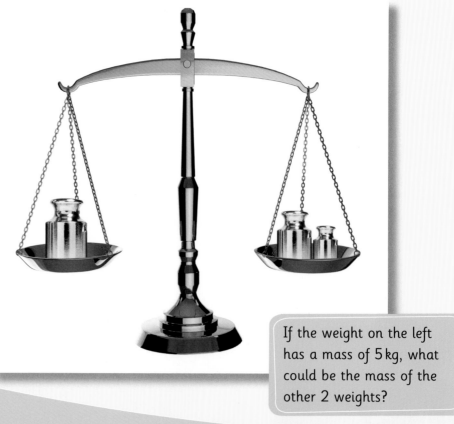

Jess swims 1 metre in 8 seconds. How long will it take her to swim the 50 m length of this pool?

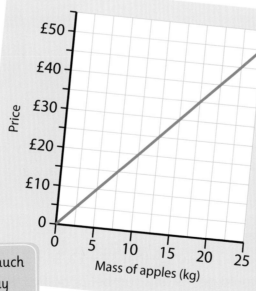

I wonder how much it will cost to buy 25 kg of apples?

Teacher's Guide

Look at the pictures with the children and discuss the questions.
See pages 140-1 of the *Teacher's Guide* for key ideas to draw out.

127 ★

You need:
- number rods
- squared paper
- rulers
- A4 paper
- scissors

Let's learn

If $a + b = 3$, then a and b must be 1 and 2!

They could be, but they could be lots of different numbers too. a could be 0.5 and b could be 1.5: a could be 5 and b could be -2.

Unknowns

Look at the bar model. It shows 4 different number statements.

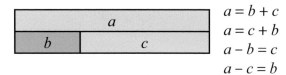

$a = b + c$
$a = c + b$
$a - b = c$
$a - c = b$

If a is 100, b and c are the unknowns.

What could b and c be?

If b is 15, what could a and c be?

Look at the pie chart.

If the whole represents 100%, what could each part be?

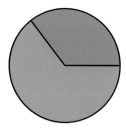

Variables

A variable is a quantity that can vary in value and size.

You know the formula for the perimeter of a rectangle: $P = 2(l + w)$ or $2l + 2w$.

Different rectangles have different lengths and widths. So l and w vary, and the value of P will vary too. P, l and w are variables.

You can use variables to write formulae.

There are approximately \$1.5 to £1. So you could say that \$3 = £2, \$6 = £4, and so on.

To work out how many pounds you would get for any number of dollars, you can use p for the number of pounds and d for the number of dollars: $d = 1.5p$

For any number of pounds you will have 1.5 times as many dollars.

Work out how many dollars you will get for £12.

Teacher's Guide

Before working through the *Textbook*, study page 142 of the *Teacher's Guide* to see how the concepts should be introduced. Read and discuss the page with the children. Provide concrete resources to support exploration.

1 Write.

Write 5 possible values for a and b in each of these.

a $\quad a + b = 48$

d $\quad b - 12 = a$

g $\quad 2a + 2b = 12$

b $\quad a + 15 = b$

e $\quad a - b = 17$

h $\quad 2b - a = 5$

c $\quad 20 + b = a$

f $\quad 2a + b = 10$

2 Convert.

If e = euro and p = pound, convert these amounts using the formula $p = 2e$.

a $\quad$ 15 euros

c $\quad$ 30 euros

e $\quad$ 18 euros

b $\quad$ £6

d $\quad$ £25

f $\quad$ £54

3 Draw.

Draw these rectangles on squared paper.

a $\quad l = 10$ cm. What could P and w be?

b $\quad l = 13$ cm. What could P and w be?

c $\quad w = 5$ cm. What could P and l be?

d $\quad w = 9$ cm. What could P and l be?

e $\quad P = 16$ cm. What could l and w be?

f $\quad P = 24$ cm. What could l and w be?

Write the formula for each of your rectangles, to show that it is correct.

P = perimeter,
l = length and
w = width.

4 Investigate.

Take a piece of A4 paper. Cut it in half widthways. Take 1 half and cut that in half. Do this again until you have 5 pieces.

Label the width of the smallest piece a and the length b.

Put the smallest piece beside the largest piece, like this.

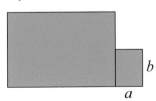

The pieces need to meet along the edges and with corners touching.

The perimeter is $10a + 4b$. Explain why.

Use pairs of rectangles to make other shapes.

Find their perimeters.

Teacher's Guide

See page 143 of the *Teacher's Guide* for ideas of how to guide practice. Work through each step together as a class to develop children's conceptual understanding.

129

Linear number sequences

Let's learn

I had to find the 50th number in this sequence: 5, 10, 15, ... It took me ages! I had to write all the numbers down.

That's one way of doing it! There is a quicker way though. We know the first number is 5 and the second is 10, so all we do is multiply the number we need by 5. So you could have calculated 50 × 5 to find your number. It's 250.

Number sequences

nth term $= 5n$

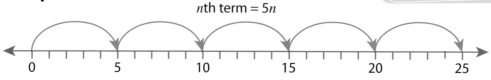

To find any number in a number sequence, you need to know the nth term.

Look at the number line. You can see that the sequence 5, 10, 15, ... goes up in fives.

The nth term is $5n$. n is where the term is in the sequence, and $5n$ gives you the term.

So for the 50th term, $n = 50$, and the term is $5 \times 50 = 250$.

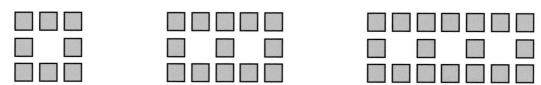

You can show number sequences using patterns. Look at the diagram above.

How many squares make up each pattern? What's the nth term for the sequence?

Linear relationships

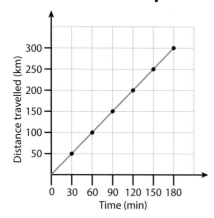

You can show the relationship between 2 variables on a graph.

This graph shows the relationship between time and distance. You can see that for every 30 minutes the distance travelled is 50 km. This is a linear relationship. The graph is a straight line.

How long does it take to travel 300 km?
What about 500 km?

Teacher's Guide

Before working through the *Textbook*, study page 144 of the *Teacher's Guide* to see how the concepts should be introduced. Read and discuss the page with the children. Provide concrete resources to support exploration.

1 Write.

Write the next 5 numbers in each of these sequences.
Write a formula for the nth term for each sequence.

a 2, 4, 6

b 3, 5, 7

c 20, 30, 40

d 1, 4, 9

e 13, 25, 37

f 28, 53, 78

g 101, 201, 301

h 1, 8, 27

2 Calculate.

Work out the nth term for each of these sequences.
Use it to find the 12th, 15th and 20th terms of each sequence.

a 4, 8, 12

b 7, 14, 21

c 4, 7, 10

d 11, 21, 31

e 8, 13, 18

f 15, 30, 45

g 14, 24, 34

h 14, 26, 38

3 Apply.

Find the mass of an apple. Use that mass to predict the masses of:

a 4 apples

b 8 apples

c 12 apples

d 16 apples

e 28 apples

f 56 apples

g 84 apples

h 100 apples

4 Think.

These patterns show a white patio and the brown flower beds that surround it.

Draw the next 5 sequences for this pattern.

Work out the formula for the nth term.
Use it to find the 10th term.

Now make up your own pattern and work out what the nth term is.

Teacher's Guide

See page 145 of the *Teacher's Guide* for ideas of how to guide practice. Work through each step together as a class to develop children's conceptual understanding.

131

Think algebra!

Let's play

Teacher's Guide

See pages 146-7 of the *Teacher's Guide*. Explain the rules for each game and allow children to choose which to play. Encourage them to challenge themselves and practise what they have learnt in the unit.

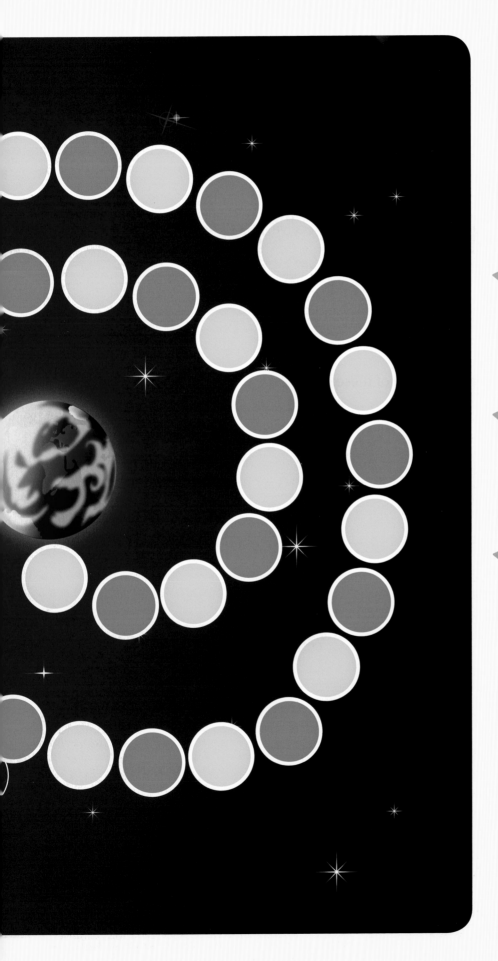

1 **Race to Earth**

Race a friend to Earth. Make up addition and subtraction statements as you go.

2 **Perimeter points**

Score points by finding the lengths and widths of rectangles with a given perimeter.

3 **Your game**

Make up your own game using the gameboard. Explain the rules and play with a partner.

Let's review

1

Sally and her friends used this formula to convert some money from pounds (p) to dollars (d):

$$d = 1.5p$$

a Sally changed £500. How many dollars did she receive?

b Freddie changed £250. How many dollars did he receive?

c Jo changed £300. How many dollars did she receive?

2

Write down 2 possibilities for s and t in each of these:

a $s + t = 36$ e $3t + 10 = s$

b $79 - t = s$ f $2s + 2t = 50$

c $156 + s = t$ g $240 - s = t$

d $2s + 4 = t$ h $158 + 2t = s$

Work out what the nth term and 10th term will be in these sequences:

i 2, 4, 6, 8   k 3, 5, 7, 9

j 5, 10, 15, 20   l 7, 12, 17, 22

Teacher's Guide

See pages 148-9 of the *Teacher's Guide* for guidance on running each task.
Observe children to identify those who have mastered concepts and those who
require further consolidation.

3

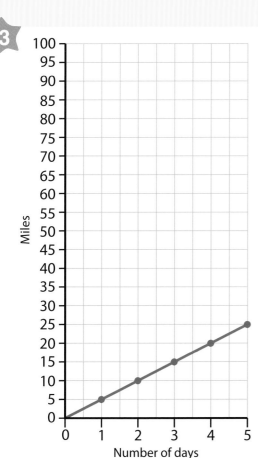

Make up 5 statements about this graph.

Now answer these questions:

a   How many miles will be travelled in 10 days?

b   How many days will it take to travel 40 miles?

c   How many days will it take to travel 100 miles?

d   How many miles will be travelled in 2 weeks?

Did you know?

Amaze your friends and family with this simple trick. Give them these instructions:
1. Pick a number, any number.
2. Multiply it by 2.
3. Add 8.
4. Divide by 2.
5. Subtract the number you started with.
Then tell them their answer is 4.

This always works.
Can you work out why?

Solving more problems

Diameter of planets

Mercury 4 878 km
Earth 12 756 km
Uranus 50 724 km
Pluto (dwarf) 2 400 km
Venus 12 104 km
Mars 6 780 km
Jupiter 139 822 km
Saturn 116 464 km
Neptune 49 248 km
SUN

The sizes of the planets are very different. How could you calculate some differences in the diameters?

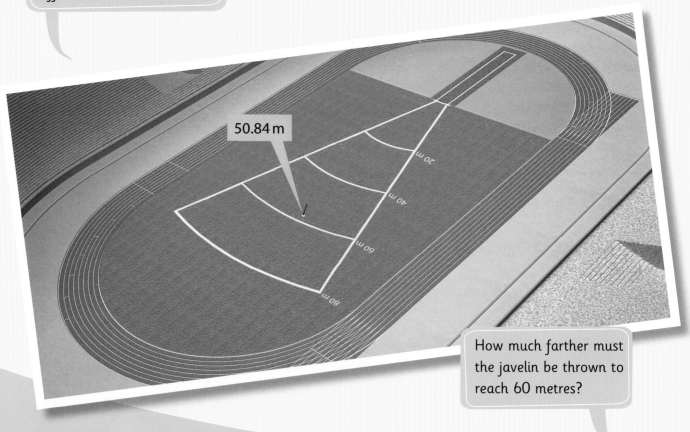

50.84 m

20 m
40 m
60 m
80 m

How much farther must the javelin be thrown to reach 60 metres?

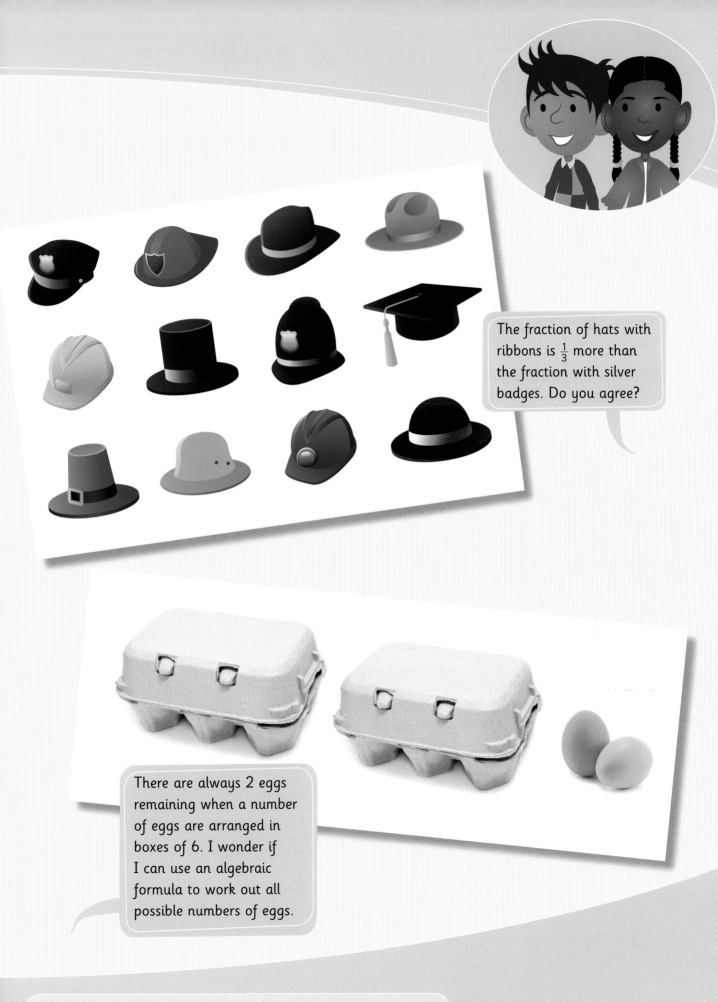

The fraction of hats with ribbons is $\frac{1}{3}$ more than the fraction with silver badges. Do you agree?

There are always 2 eggs remaining when a number of eggs are arranged in boxes of 6. I wonder if I can use an algebraic formula to work out all possible numbers of eggs.

Teacher's Guide
Look at the pictures with the children and discuss the questions.
See pages 150–1 of the *Teacher's Guide* for key ideas to draw out.

137 ★

Let's learn

I always do multiplication and division before addition and subtraction.

Yes, but remember that brackets come first in BIDMAS. You will need to do the addition or subtraction first if it is in brackets!

You need:
- pencil and paper
- number rods

Representing problems

The bar model can be used to represent and help solve problems.

Look at the problem here. What does each part show?

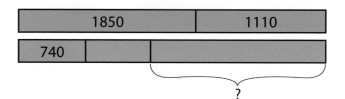

| 1850 | 1110 |

| 740 | | |
?

> In a car rally across Europe, drivers can choose to follow the green or red route.
> Both routes cover an equal distance.
> The first section of the green route is 1850 km followed by a second section of 1110 km.
> The first 2 sections of the red route are 740 km each.
> What is the distance of the final section of the red route?

The green bars represent the total distance when 1850 km and 1110 km are added together.

The red bars total the same distance but are made up of 3 parts, 2 of which are 740 km as the problem tells you that the first 2 sections are the same. The last unlabelled part is the distance you need to find to solve the problem.

Solving problems

To solve the problem, which operations are needed and in what order?

The problem can be solved in different ways. All of these equations match the problem. Match each part of the equations to the bar model to check that you agree.

- 1850 km + 1110 km = (740 km × 2) +

- ▢ = 1850 km + 1110 km − 740 km − 740 km

- ▢ = (1850 km + 1110 km) − (740 km × 2)

Teacher's Guide

Before working through the *Textbook*, study page 152 of the *Teacher's Guide* to see how the concepts should be introduced. Read and discuss the page with the children. Provide concrete resources to support exploration.

1

Calculate.

Calculate the value of the white bar.

a

?

399	598.5	399	598.5

c

1200		

1599	?

b

?

24	64.45	35.55

d

2.675	?

0.85			

2

Write.

a This equation matches the first bar model in Step 1.

$399 + 598.5 + 399 + 598.5 =$ ▢

Write the equation in a different way.

b Now write at least 1 equation to match each of the other bar models in Step 1. Remember to think about additive and multiplicative reasoning.

3

Draw.

Draw a bar model to represent and solve this problem.

Eva and I have tennis lessons on a Saturday morning lasting a total of 2 hours 15 minutes. We spend 25 minutes playing a game and the rest of the time is split equally into 2 lessons. What is the length of **1** lesson?

4

Think.

Choose a bar model from Step 1 and create your own word problem to match it.

Try to use a context of measurement, e.g. length, capacity or mass.

Now try another representation and create a word problem using a different context.

And another.

Teacher's Guide

See page 153 of the *Teacher's Guide* for ideas of how to guide practice. Work through each step together as a class to develop children's conceptual understanding.

139 ⭐

Solving problems involving fractions

Let's learn

You always need to multiply the denominators together when you want to find a common denominator.

Not always! You need to check to see if the denominators are already a multiple of the same number.

You need:
- protractor
- fraction bars

Equivalent fractions

The pie chart is divided into 3 different sectors.

The sectors show the fraction of the total number of tickets sold each week.

The blue sector represents $\frac{1}{12}$.

The red sector represents $\frac{3}{12}$ or $\frac{1}{4}$.

What fraction does the green sector represent?

Tickets sold for Sports Day

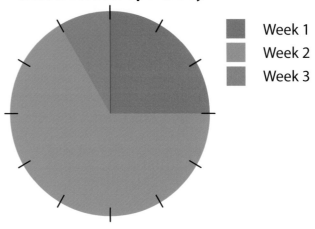

Week 1
Week 2
Week 3

You can use fraction bars to confirm that $\frac{8}{12} = \frac{2}{3}$.

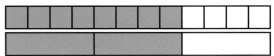

Adding and subtracting fractions

Knowing about equivalent fractions is important when adding or subtracting fractions with different denominators.

You can find the total fraction of tickets sold in week 1 and week 3 by adding $\frac{1}{4} + \frac{1}{12}$.

This can be rewritten as $\frac{3}{12} + \frac{1}{12} = \frac{4}{12}$.

$\frac{4}{12}$ can be simplified to $\frac{1}{3}$ because:

$$\frac{4}{12} \div \frac{4}{4} = \frac{1}{3}$$

Now find the fraction that shows:

- the total tickets sold in the first 2 weeks
- the difference between the tickets sold in week 2 and in week 3
- the difference between the tickets sold in week 1 and in week 2.

Teacher's Guide

Before working through the *Textbook*, study page 154 of the *Teacher's Guide* to see how the concepts should be introduced. Read and discuss the page with the children. Provide concrete resources to support exploration.

1

Explore.

a What fraction of the flag does each colour represent?

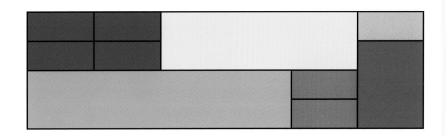

b Now explore adding the fractions to find different totals and then calculate the differences between pairs of fractions.

c What is the total fraction of green + yellow + blue?

2

Calculate.

Use equivalent fractions to calculate these:

a $1\frac{3}{8} + \frac{3}{4} =$

c $1\frac{4}{5} - \frac{7}{10} =$

e $4\frac{1}{2} + 2\frac{2}{3} =$

b $1\frac{3}{8} + \frac{1}{3} =$

d $2\frac{4}{5} - 1\frac{2}{3} =$

f $\frac{5}{6} + \frac{5}{12} + 1\frac{3}{4} =$

3

Apply.

The money raised at a local car-boot fair came from refreshments, sellers' pitches and buyers' entrance fees.

- The money raised by entrance fees and sellers' pitches represents $\frac{3}{4}$ of the total sales.
- The money raised by refreshments and entrance fees represents $\frac{5}{12}$ of the total sales.
- The difference between the money raised by entrance fees and sellers' pitches represents $\frac{5}{12}$ of the total sales.

Draw and label a pie chart to represent the fraction of total money raised from each of the above. Use a protractor to help you.

4

Think.

Eva writes down 3 fractions.

Their mean value is $\frac{1}{2}$.

Which 3 fractions did Eva write?

Investigate to find as many possible sets of solutions as you can.

Teacher's Guide

See page 155 of the *Teacher's Guide* for ideas of how to guide practice. Work through each step together as a class to develop children's conceptual understanding.

141 ★

11c Finding possible solutions for equations

You need:
- number rods
- cubes
- squared paper
- ruler

Let's learn

You can only have 1 unknown value in an algebraic equation.

That's not true. Some equations need you to find pairs of numbers to solve 2 unknowns, e.g. $12 + a + b = 18$

Using simple formulae

This line graph represents a simple formula.

You can use a table to help you look for patterns. Being systematic is very useful too!

x value	y value
2	0
3	1
4	2

You can see that, for any point on the line, the value of the x coordinate is 2 more than the value of the y coordinate.

The formula is $x = y + 2$ or $y = x - 2$.

x and y are both variables because their values vary.

What is the value of y when x has the value 12.5?

Finding possible solutions

Eva doubles a number and then adds 4. What could her starting number have been? What is her answer?

The number you need to double is unknown so you can represent it with a bar of any length.

Then you know that 4 is added to the doubled number to reach a total.

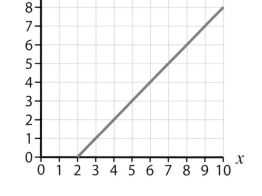

Here, 3 is used to show the unknown number or variable n, so a possible solution is the pair of numbers 3 and 10.

You can write the formula $2n + 4$ to help you find other possible solutions.

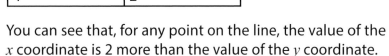

Teacher's Guide

Before working through the *Textbook*, study page 156 of the *Teacher's Guide* to see how the concepts should be introduced. Read and discuss the page with the children. Provide concrete resources to support exploration.

1

Find.

Find the values of d and n.

Ali uses cubes to represent the formula $d = n^2 - 2$.

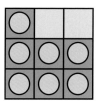

a   What is the value of n and d in Ali's representation?

b   Find other pairs of values for n and d using the same formula.

2

Calculate.

Use the bar model to represent the formula $2n + 4 = m$ when n has the following values:

a   6   c   12   e   23

b   10   d   16   f   27

Write down the values of n and m each time.

3

Draw.

Ali is planning a party. He would like each of his friends to have 2 small presents to take home, but he wants an extra present for the winner of a game.

He writes a formula to help him work out how many presents he needs to buy in total.

What formula should Ali use?

Draw a graph using Ali's formula and use a table to list the number of friends and presents each time.

4

Investigate.

The area of these shapes is expressed algebraically as $A = 3n - 3$.

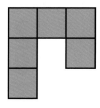 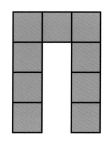

Investigate the area of other shapes that suit the expression $A = 3n - 3$.

Write down the values for A and n each time.

Teacher's Guide

See page 157 of the *Teacher's Guide* for ideas of how to guide practice. Work through each step together as a class to develop children's conceptual understanding.

143

Fraction frenzy!

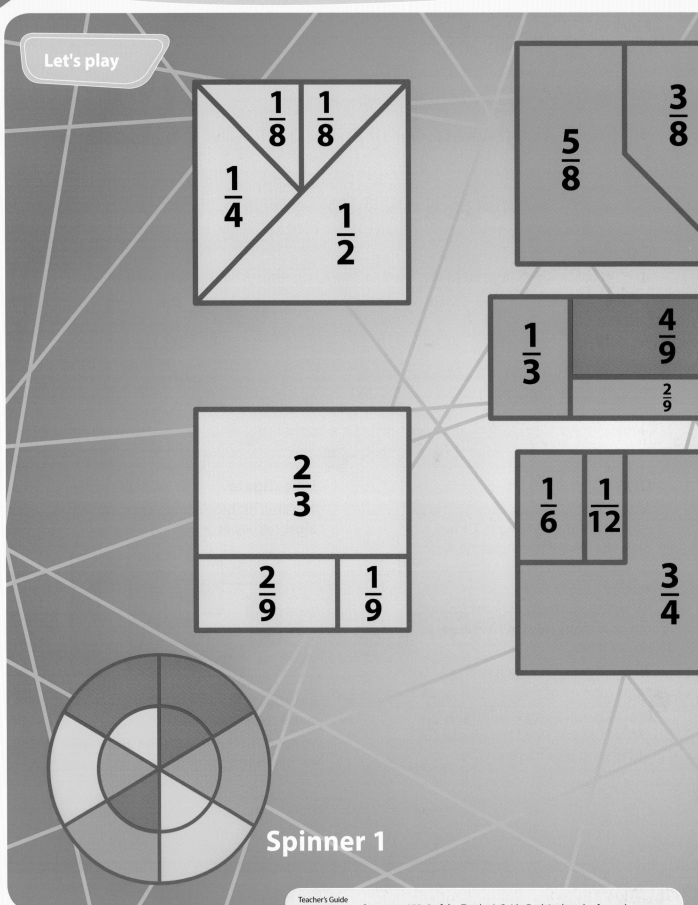

Let's play

$\dfrac{1}{8}$ $\dfrac{1}{8}$

$\dfrac{1}{4}$

$\dfrac{1}{2}$

$\dfrac{5}{8}$ $\dfrac{3}{8}$

$\dfrac{1}{3}$ $\dfrac{4}{9}$

$\dfrac{2}{9}$

$\dfrac{2}{3}$

$\dfrac{2}{9}$ $\dfrac{1}{9}$

$\dfrac{1}{6}$ $\dfrac{1}{12}$

$\dfrac{3}{4}$

Spinner 1

Teacher's Guide

See pages 158–9 of the *Teacher's Guide*. Explain the rules for each game and allow children to choose which to play. Encourage them to challenge themselves and practise what they have learnt in the unit.

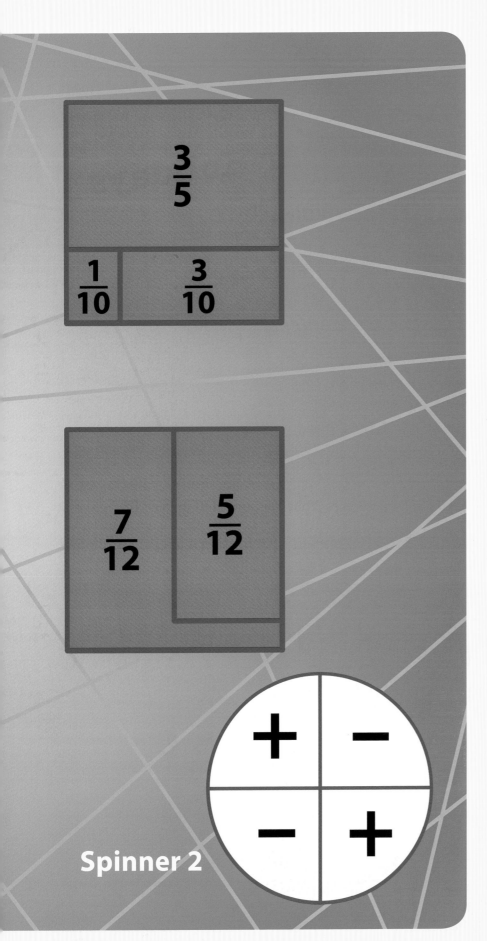

$$\frac{3}{5}$$

$$\frac{1}{10} \quad \frac{3}{10}$$

$$\frac{7}{12} \quad \frac{5}{12}$$

Spinner 2

 Making calculations

Make up addition and subtraction calculations using the fractions on the gameboard.

 The answer is ...

Create calculations that give the fractions shown on the gameboard.

 Your game

Make up your own game using the gameboard.

And finally ...

Let's review

1

a The total cost of a holiday for a family with 2 adults and 3 children is £1017.
The cost for 1 child is £149.
What is the cost for 1 adult?

b Now use the prices to work out the cost of 3 adults and 4 children.

Star Holidays

Invoice _____

3 × children @ £149

2 × adult @

2

a Eva pours squash from glass B to fill up glass A. What fraction is left in glass B?

b Ali pours more squash from glass B to fill up glass C. What fraction is left in glass B now?

$\frac{5}{6}$

A

$\frac{2}{3}$

B

$\frac{3}{5}$

C

Teacher's Guide

See pages 160–1 of the *Teacher's Guide* for guidance on running each task.
Observe children to identify those who have mastered concepts and those who require further consolidation.

3

The site manager wants to work out the number of chairs and tables that are needed for different events. Tables and chairs are always arranged like this, but 3 additional chairs are always placed near the fire exit.

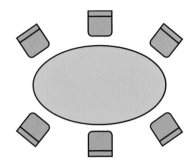

a Write an algebraic equation that the site manager can use to help him with his problem, where n is the number of tables each time.

b Now use the equation to complete the solutions below:

Tables	3		9	13		20		28
Chairs		33			93		153	

Did you know?

From as early as 1800 BCE, the Egyptians were writing fractions. Their number system was a Base 10 idea (a little bit like ours now) so they had separate symbols for 1, 10, 1000, 10 000, 100 000 and 1 000 000.

The Egyptians wrote all their fractions using unit fractions. They drew a mouth picture to represent 'part' above a number to make it into a unit fraction. So $\frac{1}{3}$ would look like this:

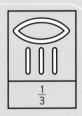

The ancient Egyptians didn't write fractions with a numerator greater than 1! They wrote fractions like $\frac{3}{4}$ as a sum of different unit fractions, so $\frac{3}{4} = \frac{1}{2} + \frac{1}{4}$ not $\frac{1}{4} + \frac{1}{4} + \frac{1}{4}$.

1	10	100	1000	10 000	100 000	1 000 000

Fractions, equivalents and algebra

I wonder how tall these trees could be?

What fraction of the coins is made of copper?

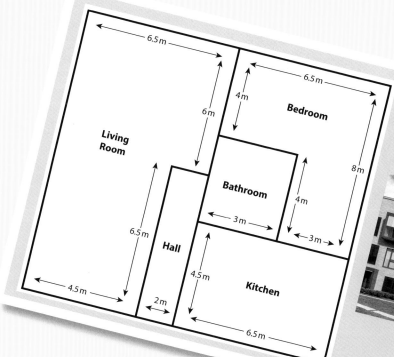

New spacious apartments
Dimensions shown

How could you find the total area of the apartment?

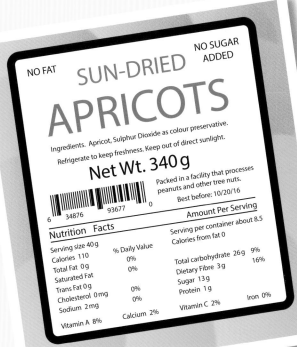

I wonder how much of this whole packet is carbohydrate?

Teacher's Guide

Look at the pictures with the children and discuss the questions.
See pages 162–3 of the *Teacher's Guide* for key ideas to draw out.

149 ★

Equivalences

Let's learn

$\frac{2}{5}$ of 40p and $\frac{8}{10}$ of 20p are both 16p so $\frac{2}{5}$ and $\frac{8}{10}$ must be equivalent fractions!

No, it doesn't quite work like that. You are finding fractions of different amounts. The answer for both is 16p but $\frac{2}{5}$ is equivalent to $\frac{4}{10}$ not $\frac{8}{10}$.

Equivalent fractions

You can use bar models to find out how many tenths are equivalent to fifths.

You can also find equivalent fractions by multiplying (or dividing) the numerator and the denominator by the same number.

Choose a fraction and make up 10 equivalent fractions using multiplication.

40p

20p

Types of fractions

Decimals and percentages are types of fractions.

Look at the pie chart. You can describe each part in 3 different ways:

$\frac{1}{4}$ is equivalent to 25% and 0.25.

- The pink part is $\frac{1}{4}$ of the pie chart.
- The pink part is 25% of the pie chart.
- The pink part is 0.25 of the pie chart.

The whole pie chart shows the favourite colours of 80 people.

How many like red best? What about yellow and white?

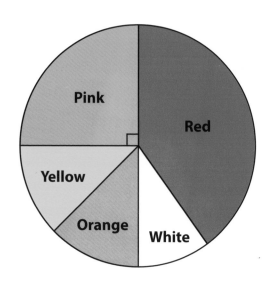

1

Find.

Write down 6 equivalent fractions for each of these:

a $\frac{1}{2}$ b $\frac{3}{4}$ c $\frac{2}{3}$ d $\frac{8}{12}$ e $\frac{3}{5}$ f $\frac{15}{25}$

2

Convert.

Write each decimal as a fraction and a percentage.

a 0.2 b 0.8 c 0.15 d 0.65 e 0.125 f 0.875

3

Solve.

Dana was offered these amounts of money:

- 15% of £300
- 0.25 of £280
- $\frac{7}{8}$ of £128

She was saving for a new bike so wanted as much money as possible.

Which offer should she go for?

Make up a similar problem and solve it.

4

Investigate.

Work with a partner.

Count the cubes in the bag your teacher has given you.

Work out what fraction each colour is of the whole amount.

Draw a pie chart to show this information.

You must label each part with a fraction or percentage. You must also give your pie chart a title which includes the total number of cubes.

When you have done this, swap your pie chart with another pair's. Then work out the number of each colour cube they had.

Compare your 2 sets of data. How are they the same? How are they different?

Teacher's Guide

See page 165 of the *Teacher's Guide* for ideas of how to guide practice. Work through each step together as a class to develop children's conceptual understanding.

151 ★

12b Formulae and sequences

Let's learn

I think that all shapes that have the same area will have the same perimeter.

I don't think you are right. You can have a rectangle that has an area of 24 cm² and a perimeter of 50 cm, another can have a perimeter of 28 cm and another can have a perimeter of 22 cm. There are several different perimeters for just 1 area!

Formulae

[Shapes: rectangle 3 cm by 8 cm; rectangle 4 cm by 6 cm; triangle 5 cm by 4 cm; triangle 7 cm; parallelogram 4 cm by 6 cm]

You can use formulae to find the areas and perimeters of different shapes.

> The area of any rectangle is $l \times w$.

To find the area of a rectangle you need to know its length and width.
You can substitute the values for the letters l and w.

> The area of any triangle is $\frac{1}{2}(b \times h)$.

> The area of a parallelogram is $l \times w$.

Work out the areas of the shapes above.

Linear number sequences

Linear number sequences increase or decrease by the same number each time.

What is the sequence shown in the 100 square?
What will the 120th number in the sequence be?

What is the nth term?

Put counters on the squares that show the sequence $3n + 1$.

1	2	3	4	5	6	7	8	9	10
11	12	13	14	15	16	17	18	19	20
21	22	23	24	25	26	27	28	29	30
31	32	33	34	35	36	37	38	39	40
41	42	43	44	45	46	47	48	49	50
51	52	53	54	55	56	57	58	59	60
61	62	63	64	65	66	67	68	69	70
71	72	73	74	75	76	77	78	79	80
81	82	83	84	85	86	87	88	89	90
91	92	93	94	95	96	97	98	99	100

1

Calculate.

Calculate the perimeter and area of rectangles with these dimensions:

a
3 cm
4 cm

c
3 cm
7 cm

b
4 cm
10 cm

d $l = 5$ cm, $w = 2$ cm

e $l = 3$ cm, $w = 1.5$ cm

f $l = 6.5$ cm, $w = 3.5$ cm

2

Write.

Write the first 10 numbers in each of these sequences:

a $2n$

b $4n + 1$

c $6n + 1$

d $3n + 2$

e $5n + 4$

f $10n + 9$

3

Draw.

Draw 5 different triangles on centimetre-squared paper.

Use the formula to find their areas.

Then find their perimeters.

Now do the same for 5 parallelograms.

4

Think.

Farmer Giles has 24 m of fencing.

Find all the possible rectangular enclosures with whole metre sides that he can make.

He wants to cover his enclosure with grass. Which one will need the least grass?

Teacher's Guide

See page 167 of the *Teacher's Guide* for ideas of how to guide practice. Work through each step together as a class to develop children's conceptual understanding.

153

Unknowns

You need:
- strips of paper
- scissors
- number rods
- counters

Let's learn

You have to use inverse operations to find the missing number in the statement 356 − n = 167. That's the only way!

It's not the only way. You can also use bar models. They help you visualise exactly what you need to do.

Two unknowns

m	
n	p

The bar model shows the relationships between the variables m, n and p.

What number statements can you make using m, n and p?

If p = 12, what could m and n be?

Missing number problems

356	
167	n

This bar model shows the relationship between the numbers in the statement 356 − n = 167.

What is the value of n?

You can use algebra to represent the unknown value you need to find in a number problem.

This bar model represents the following problem:

A shop had a sale. 20% was taken off the price of a TV. The sale price is £240. What was the original price?

£x				
20%	20%	20%	20%	20%

£240

What is £x?

Teacher's Guide

Before working through the *Textbook*, study page 168 of the *Teacher's Guide* to see how the concepts should be introduced. Read and discuss the page with the children. Provide concrete resources to support exploration.

1

Answer these.

Solve these missing number statements:

a $156 + n = 267$

b $m + 258 = 574$

c $467 + s = 984$

d $536 - t = 319$

e $m - 1325 = 1284$

f $n - 2489 = 3573$

2

Answer these.

Solve these missing number statements:

a $m + n = 12.4$ d $21.3 + w = y$

b $15.6 - s = t$ e $m + 13.2 = b$

c $12.8 + r = v$ f $c - 12.7 = d$

3

Solve.

Solve these missing number problems:

a   Ben had 12 books. Sam has 4 times as many. How many more books did Sam have? How many books did they have altogether?

b   A pair of jeans has been reduced by 20% in a sale. They are now £24. How much were they before the sale? They are reduced by a further 10%. What is the price now?

c   Polly ran 12.7 km from her home to the shop. She then ran back 6.8 km and stopped to talk to a friend. How far was she from home?

d   Sasha collected 356 shells. She gave some to a friend and kept 167. How many did she give to her friend?

e   Parveen picked 2 red apples for every 3 green apples. She picked 60 red apples. How many green apples did she pick?

f   Nina made a fruit salad. She put 1.5 kg of banana into it. She put in a quarter of that amount of cherries. What mass of fruit did she use?

4

Think.

Josie had 7 times as many sweets as Abi.

Josie gave Abi some of her sweets. They now each have the same amount.

How many sweets did Josie have before sharing them with Abi? Write down 6 possibilities. Make up a similar problem and solve it.

Teacher's Guide

See page 169 of the *Teacher's Guide* for ideas of how to guide practice. Work through each step together as a class to develop children's conceptual understanding.

155

Odd and even four in a row

Let's play

E	O	O	E
E	O	O	E
E	O	E	O
O	E	O	E
O	E	O	E
O	E	E	O

Teacher's Guide

See pages 170–1 of the *Teacher's Guide*. Explain the rules for each game and allow children to choose which to play. Encourage them to challenge themselves and practise what they have learnt in the unit.

1 **Find the fifth term**

Make sequences. Is the fifth term odd or even?

2 **Discount dice**

Find the the original price if the price given has a 20% discount. Is the answer odd or even?

3 **Your game**

Make up your own game using the gameboard.

Let's review

1

Look at these fractions. Write 2 more fractions for each that are equivalent.

Now make them equivalent to a decimal and percentage.

a $\frac{2}{5}$ c $\frac{3}{4}$ e $\frac{9}{10}$ g $\frac{3}{8}$

b $\frac{1}{10}$ d $\frac{1}{2}$ d $\frac{4}{5}$ h $\frac{5}{8}$

Next, write a list of 8 decimals that you have not used yet and write them as fractions and percentages.

Finally, write a list of 8 percentages that you have not used yet and write them as fractions and decimals.

2

Complete these linear number sequences and then write the formula for the 10th term:

a   12, 24, 36, ___ , ___ , ___

b   25, 50, 75, ___ , ___ , ___

c   5, 9, 13, ___ , ___ , ___

d   7, 12, 17, ___ , ___ , ___

e   16, 31, 46, ___ , ___ , ___

f   55, 105, 155, ___ , ___ , ___

Teacher's Guide

See pages 172–3 of the *Teacher's Guide* for guidance on running each task.
Observe children to identify those who have mastered concepts and those who require further consolidation.

3

Draw each problem using the bar model, then write an algebraic statement for each and solve them:

a   Sam had some marbles. He gave 26 to his friend and was left with 48.
How many marbles did he start off with?

b   Adam had 3 times as many football stickers as Tom. Tom had 45.
How many did Adam have?

c   Rosie baked 75 muffins. She gave 36 to her friend. How many did she have left?

d   Samira ate 3 times as many cherries as Sally. Samira ate 27 cherries.
How many cherries did Sally eat?

What could the values of these letters be? Make up 5 possibilities for each question.

e $n + m = 145$

f $1568 - s = t$

g $34\,510 + k = h$

h $t - w = 2345$

i $a + 3568 = b$

j $d - z = 32\,567$

Did you know?

The word 'algebra' came from the Arabic word *al-jabr*.

Algebra was first used in ancient Egypt and Babylon. The Persian mathematician Muhammad ibn Musa Al-Khwarizmi is credited as one of the forefathers of algebra. This Russian stamp commemorates 1200 years since his birth.

Fair shares

How much would half each quantity be?

How many buses would be needed to take everyone on the trip?

I wonder what day of the week my birthday will be on this year?

SUN	MON	TUE	WED	THU	FRI	SAT
27	28	29	30	1	2	3
4	5	6	7	8	9	10
11	12	13	14	15	16	17
18	19	20	21	22	23	24
25	26	27	28	29	30	31

Carrots 83p per kg

If I bought 2 kg of carrots and 3 swedes for £3.01, how much would that make the swedes each?

I would love a class trip to the seaside! I wonder how much it would cost us each if we hired a coach?

Teacher's Guide

Look at the pictures with the children and discuss the questions.
See pages 174–5 of the *Teacher's Guide* for key ideas to draw out.

161 ⭐

Using long division

Let's learn

I worked out that my class needs 2 minibuses to go swimming. There are 29 children and each minibus holds 12 children. 29 ÷ 12 = 2R5 so 2 is the answer.

That won't work – 5 children will be left behind! You need to round up so that everybody can go swimming. You need 3 minibuses.

Long division

Look at 2921 ÷ 23. How many groups of 23 can you make from 2921?

```
        127
   23) 2921
       23
        6
```

```
        127
   23) 2921
       23
       62
       46
       16
```

```
        127
   23) 2921
       23
       62
       46
      161
      161
```

```
        127
   23) 2921
       23
       62
       46
      161
      161
```

2000 cannot be grouped into 2300s

2900 can be grouped into 1 lot of 2300 with 600 left over

600 is combined with 20 to make 620

620 can be grouped into 2 lots of 230

$2 \times 230 = 460$ with 160 left over

160 is combined with 1 to make 161

161 can be grouped into 7 lots of 23

$7 \times 23 = 161$ so there is no remainder

```
         1 2 7
   23) 2²9⁶2¹⁶1
```

```
         1 2 7
   23) 2²9⁶2¹⁶1
```

```
         1 2 7
   23) 2²9⁶2¹⁶1
```

```
         1 2 7
   23) 2²9⁶2¹⁶1
```

Dealing with remainders

Think about 44 ÷ 7 = 6R2
The answer may be:

| 6 | 7 | $6\frac{2}{7}$ | 6.286 to 3 decimal places | 6 remainder 2 |

You have to consider the context to decide which is correct.

1

Answer these.

Give any remainders as fractions and decimals.

a $3471 \div 13$

b $5066 \div 34$

c $6612 \div 76$

d $586 \div 8$

e $87 \div 7$

f $876 \div 13$

2

Copy and complete.

a $305 = \boxed{} \div 26$

b $56 \times \boxed{} = 2072$

c $4\boxed{} \times \boxed{}6 = 3698$

Which are correct? Explain why each one is correct or incorrect.

d $88 \div 3 = 29.1$

e $88 \div 3 = 29\frac{1}{3}$

f $88 \div 3 = 29R1$

g $88 \div 3 = 29.3$

h $88 \div 3 = 29R3$

i $469 \div 4 = 117\frac{1}{4}$

j $469 \div 4 = 117\frac{3}{4}$

k $469 \div 4 = 117R1$

l $352 \div 3 = 117R1$

3

Apply.

Answer these word problems using long division.
Decide how to treat the remainder.

a Ali takes 13 minutes to prepare a pizza. How many can he make in a 4 hour shift?

b The circumference of a bicycle wheel is 94 cm. How many times does the wheel turn to travel 50 m?

c A box holds 54 books. How many boxes do you need to hold 3000 books?

d There are 24 children in a class. They share 100 cup cakes equally. How many do they have each?

e Eva works out the mean time for running 400 m. It is 1.3 minutes. What is that in minutes and seconds?

4

Think.

I'm thinking of a number. I divide it by 4. The remainder is 1.

What numbers could Ali have been thinking of?

Describe them. What if the remainder is 2?

What do you notice about these numbers? What if the remainder was 3? Or 4?

Teacher's Guide

See page 177 of the *Teacher's Guide* for ideas of how to guide practice. Work through each step together as a class to develop children's conceptual understanding.

163 ★

Choosing operations to solve problems

You need:
- number rods
- coins

Let's learn

I solved this word problem: Eva has £1.80 and Ali has £2.75. How much more does Ali have than Eva? The answer is £4.55. I know that because it says 'more' in the question, you have to add the numbers.

No, the answer is 95 pence! It asks how much more Ali has than Eva. You need to subtract £1.80 from £2.75 to find the difference. You have to think about the relationship between the numbers.

One operation

The bar model shows 2 amounts that together are equivalent to a third amount.

It can model addition, with the top bar being the sum.

It can model subtraction with either of the bottom bars being the difference.

It can model multiplication, with the top bar being the product.

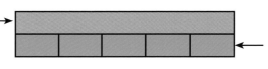

It can model division, with any of the bottom bars being the quotient or divisor.

This bar model shows 5 equal bars that together are equivalent to another amount.

5 is the multiplier and the divisor.

More than one operation

Each bar represents several possible calculations. The relationship between the quantities is the same for each of those calculations.

Notice that:
- multiplication and division diagrams have repeats of the same amount
- the longest bar represents a total or whole.

Multiply and add or subtract and divide?

Divide and there is a remainder or multiply and add?

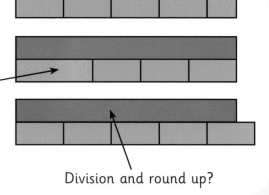

Division and round up?

Teacher's Guide

Before working through the *Textbook*, study page 178 of the *Teacher's Guide* to see how the concepts should be introduced. Read and discuss the page with the children. Provide concrete resources to support exploration.

★ 164

1

Answer these.

Identify the calculation to be done for each problem by drawing a bar model.

a   Eva has 3 pens. Ali has 3 times as many. How many pens does Ali have?

b   Eva has 6 more pens than Ali. Ali has 20 pens. How many pens does Eva have?

c   A group of children have 5 pens each. 6 children are in the group. How many pens are there?

d   Some children have 2 pens each. There are 18 pens altogether. How many children are there?

e   The average number of pens per child in a class is 5. That is 2 more than Eva has. How many pens does Eva have?

2

Answer these.

Pens cost 12p each and rulers cost 30p each.

a   Eva has £3. She buys 7 rulers. How many pens can she afford?

b   Ali buys 4 rulers and 3 pens. What is the cost?

c   Eva buys 6 rulers and 11 pens. How much change does she get from £5?

d   Ali has £3. He buys 6 rulers and 8 erasers. How much does each eraser cost?

3

Solve.

Explain why you chose the operations that you did.

a   There are 6 chairs in each row. There are 7 rows. How many chairs are there? 60 people need seats. How many more chairs would you need? How could you arrange the chairs to keep the rectangular shape?

b   Ali has 120 g of fudge. Each piece weighs 8 g. How many pieces of fudge are there? If fudge costs £2.37 per 100 g, how much will Ali's fudge cost?

c   Eva has 20 egg boxes. They each hold 6 eggs. Eva has 89 eggs and fills egg boxes with them. How many boxes remain empty?

d   Ali has 3 litres of juice. He fills 125 ml glasses with the juice. How many glasses are filled?

4

Think.

Here are some bar models for calculations.

Write 2 word problems for each one.

23	
7	16

30				
6	6	6	6	6

100		
35	35	30

Write a more complicated word problem that has more steps.

Teacher's Guide

See page 179 of the *Teacher's Guide* for ideas of how to guide practice. Work through each step together as a class to develop children's conceptual understanding.

165 ★

Multiplying and dividing fractions

You need:
- counters
- whiteboards

Let's learn

I multiplied $\frac{2}{3} \times \frac{1}{4}$ and my answer is $\frac{24}{12}$. I did it by multiplying 2 and 4 and then 3 and 1. I multiplied my answers to get 24. I then multiplied 3 and 4 to get 12.

That can't be right. $\frac{2}{3}$ of $\frac{1}{4}$ will be smaller than $\frac{1}{4}$. $\frac{24}{12}$ is 2 which is much larger! The denominator is 12, but the numerator is 2. So you get $\frac{2}{12}$ which simplifies to $\frac{1}{5}$.

Multiplying fractions

Look at $\frac{1}{3} \times \frac{1}{4}$

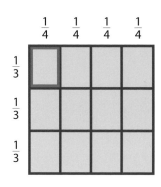

The 1 by 1 square is split into quarters along 1 side and thirds along the other side.

Each part is $\frac{1}{12}$.

$$\frac{1}{3} \times \frac{1}{4} = \frac{1}{12}$$

Look at this bar.
It is 1 unit long and 1 section is $\frac{1}{3}$.

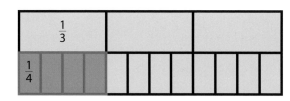

The multiplier is $\frac{1}{4}$ so the multiplicand is split into 4 equal parts.

Each part is $\frac{1}{12}$ of the whole.

$$\frac{1}{3} \times \frac{1}{4} = \frac{1}{12}$$

Dividing fractions by a whole number

Look at $\frac{1}{3} \div 5$

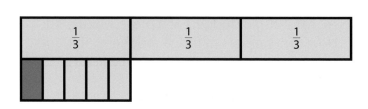

The bar represents 1.

The dividend is $\frac{1}{3}$. The divisor is 5 so $\frac{1}{3}$ is split into 5 equal parts.

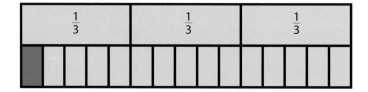

There are three $\frac{1}{3}$s so the shaded part is $\frac{1}{15}$ of the whole.

Teacher's Guide

Before working through the *Textbook*, study page 180 of the *Teacher's Guide* to see how the concepts should be introduced. Read and discuss the page with the children. Provide concrete resources to support exploration.

1

Calculate.

a $\frac{1}{2} \times \frac{1}{4}$

c $\frac{1}{3} \times \frac{1}{4}$

e $\frac{1}{5} \div 2$

g $\frac{3}{5} \div 4$

b $\frac{1}{2} \times \frac{3}{4}$

d $\frac{2}{3} \times \frac{3}{4}$

f $\frac{1}{6} \div 4$

h $\frac{6}{7} \div 3$

2

Copy and complete.

a $\frac{1}{\blacksquare} \times \frac{\blacksquare}{5} = \frac{1}{15}$

c $\frac{\blacksquare}{15} = \frac{1}{\blacksquare} \times \frac{2}{3}$

e $\frac{1}{4} = \frac{1}{2} \div \blacksquare$

g $\frac{\blacksquare}{\blacksquare} \div 5 = \frac{2}{15}$

b $\frac{2}{\blacksquare} \times \frac{\blacksquare}{7} = \frac{8}{21}$

d $\frac{9}{\blacksquare} = \frac{\blacksquare}{4} \times \frac{3}{7}$

f $\frac{\blacksquare}{\blacksquare} \div 7 = \frac{1}{49}$

h $\frac{\blacksquare}{18} \div \frac{5}{9} = \blacksquare$

3

Solve.

Ali makes potato salad using this recipe.

a For 6 people.
How much sour cream does he use?

b For 4 people.
How much chopped chives does he use?

c For 8 people.
What weight of potatoes does he use?

d Using 2 spring onions.
How many people does it serve?

e Using 1 cup of mayonnaise.
How much chopped parsley does he use?

f Eva reads that it takes $\frac{3}{4}$ of an hour to make the recipe. She estimates that she can do it in $\frac{2}{3}$ of the time. How long does she think it will take?

> **Potato salad, serves 12**
> $1\frac{1}{2}$ kg new potatoes
> $\frac{1}{2}$ cup mayonnaise
> $\frac{1}{4}$ cup sour cream
> $\frac{1}{3}$ cup chopped chives
> $\frac{2}{5}$ cup chopped parsley
> 8 spring onions

4

Think.

| $\frac{1}{2}$ | $\frac{1}{3}$ | $\frac{1}{4}$ | 12 |

Make each of the numbers from 1 to 10 using the 4 operations and the numbers in the box no more than once each.

You do not have to use all of the numbers.

Teacher's Guide
See page 181 of the *Teacher's Guide* for ideas of how to guide practice. Work through each step together as a class to develop children's conceptual understanding.

167 ★

Challenging times

Let's play

Challenge 1

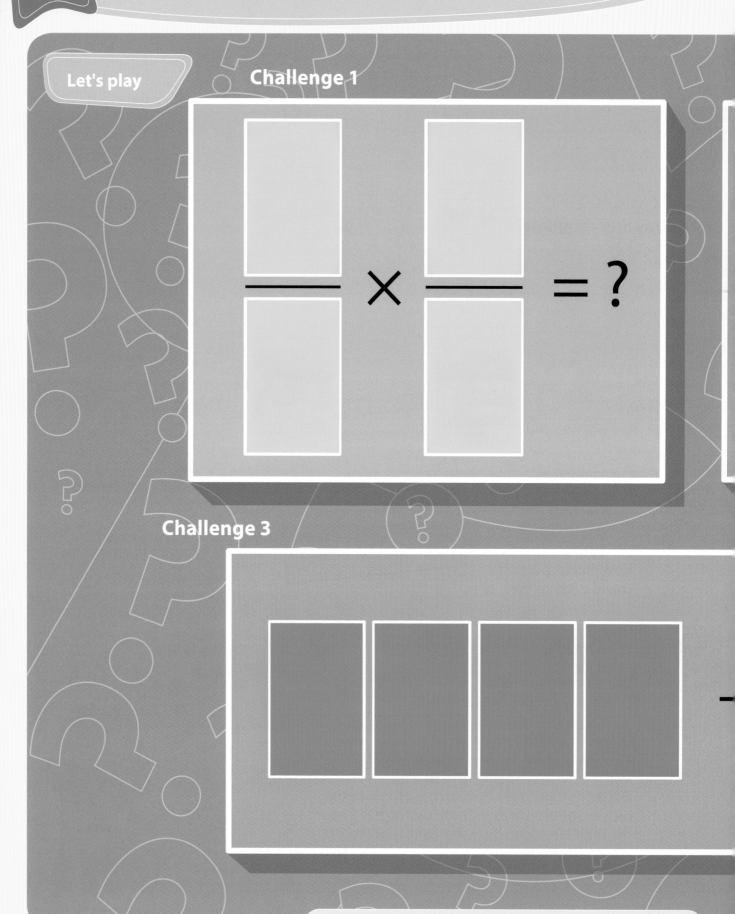

$$\overline{} \times \overline{} = ?$$

Challenge 3

Teacher's Guide

See pages 182–3 of the *Teacher's Guide*. Explain the rules for each game and allow children to choose which to play. Encourage them to challenge themselves and practise what they have learnt in the unit.

Challenge 2

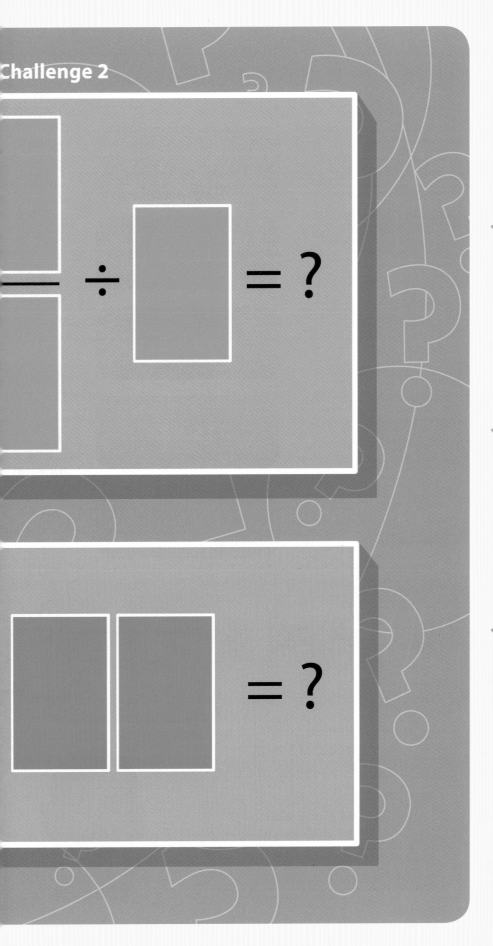

1 **Can you do it?**

Rise to the challenge and answer the calculation of your choice. Pick digit cards to fill in the blanks. Who will reach a score of 11 first?

2 **Aim high**

Now do the same but this time aim to get the highest answer that you can! Which challenge will you pick and who will be the winner with 10 points?

3 **Your game**

Design your own game using the gameboard. Explain the rules and play with a partner.

Let's review

1

Match each problem to a calculation.

Explain your answer.

You may use a diagram.

I have 15 pizzas. I cut each one into 3 pieces. How many pieces?	15 + 3
I have 15 pizzas. I eat 3 of them. How many are there now?	15 − 3
I have 15 pizzas. 2 friends share them with me. How many do we eat each?	15 × 3
I have 15 pizzas. Earlier I ate 3 of them. How many did I have before that?	15 ÷ 3
I have 15 pizzas. Each costs £3 and there is a £3 delivery charge. What do I pay?	15 × 3 + 3

2

Sometimes when I divide 2 numbers I get a decimal that goes on and on. The decimal part repeats after a while. How you can tell when it is going to start repeating?

You need:
● calculator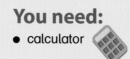

$10 \div 3 = 3.333\ldots$ has a repeat of 1 digit.

$34 \div 6 = 5.6666\ldots$ has a repeat of 1 digit.

$17 \div 7 = 2.428571428\ldots$ has a repeat of 6 digits.

What happens if you work out $46 \div 7$? Or $19 \div 6$? Investigate different divisors and see what you notice about the remainders.

Teacher's Guide

See pages 184–5 of the *Teacher's Guide* for guidance on running each task. Observe children to identify those who have mastered concepts and those who require further consolidation.

3

Choose 3 digit cards, e.g. 1, 2 and 3.
Replace the boxes below with the cards.

You need:
- 1–9 digit cards

$$\frac{\Box}{\Box} \div \Box$$

Work out the answer.
Now try the same digits but in a different order. Work out that answer.
Repeat for all 6 different arrangements of the 3 digits.

What if you use 4 cards and place them in these boxes.
What different answers can you make?

$$\frac{\Box}{\Box} \times \frac{\Box}{\Box}$$

What do you notice? Can you explain why that happens?

Did you know?

Fractions were used a long time before decimals were 'invented'. The ancient Greeks thought that all parts of a whole number could be written as fractions. We call these numbers rational numbers today.

There are also some numbers called irrational numbers. These can only be written as decimals with an infinite number of decimal places and no repeating patterns. When the ancient Greeks came across these they thought they were evil!

171

Nets, angles and coordinates

What 3-D shapes can you see?

I wonder how these models compare in size to real life?

What angles can you see?

What 2-D shapes can you see?

Can you describe the reflections?

Teacher's Guide

Look at the pictures with the children and discuss the questions.
See pages 186–7 of the *Teacher's Guide* for key ideas to draw out.

173 ★

Making and measuring 3-D shapes

Let's learn

I think
1 m³ equals
10 000 cm³.

No, it's a bigger number than that! Let's work it out.
1 m³ is 1 m × 1 m × 1 m which is 100 cm × 100 cm × 100 cm so it's 1 000 000 cm³ – that's a million!

Nets of pyramids

The shape of the base defines a pyramid.

Here are 2 nets for a pentagonal-based pyramid.

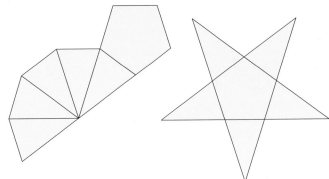

As the number of sides on the base increases, the base of each triangle decreases in length. The shape becomes closer and closer to a cone.

Converting between units of volume

Volume is measured in cubic units: mm³, cm³, m³ and km³.

1 cm³ = 1 cm × 1 cm × 1 cm = 10 mm × 10 mm × 10 mm = 1000 mm³

To convert mm³ to cm³ divide by 1000.

To convert cm³ to mm³ multiply by 1000.

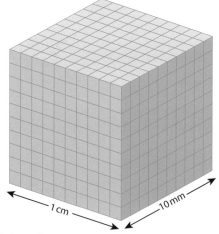

1 cm   10 mm

1 m³ = 1 m × 1 m × 1 m = 100 cm × 100 cm × 100 cm = 1 000 000 cm³

To convert cm³ to m³ divide by 1 000 000.

To convert m³ to cm³ multiply by 1 000 000.

1 km³ is so huge that it is not really a practical unit!

1 km³ = 1 km × 1 km × 1 km = 1000 m × 1000 m × 1000 m = 1 000 000 000 m³

That's a thousand million metres cubed!

Teacher's Guide

Before working through the *Textbook*, study page 188 of the *Teacher's Guide* to see how the concepts should be introduced. Read and discuss the page with the children. Provide concrete resources to support exploration.

1 Make.

Make a net for a hexagonal-based pyramid.

- Use compasses to draw a circle of radius 5 cm on thin card.
- Mark the circumference with the radius to give a regular hexagon.
- Use a ruler and protractor to draw isosceles triangles with base angles equal to 75° on each side of the hexagon to make the net.

- Add tabs to 1 side of each triangle for fastening.
- Cut out the shape, crease the fold lines and glue it together.

2 Answer these.

a   Look at these 2 cuboids.
    Which has the greater volume?

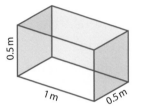 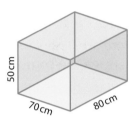

b   A jeweller has bought 1000 tiny cubes of gold that measure 125 mm³. What are the dimensions of each cube? He packs them in boxes that are 2 cm × 2 cm × 10 cm. How many boxes does he need to pack them all?

3 Apply.

A biscuit manufacturer is designing new packaging. Here are the designs:

a   b 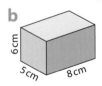   c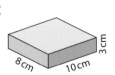

Calculate the volume of each box. What do you notice?

Work out another possible set of measurements.

Sketch the nets for all 4 boxes.

Calculate the total area of the 6 faces for each box. (You could record these in a table).

Make the box you think is best. Explain why.

4 Investigate.

The red cubes are composed of 1 m³ cubes.

Ali wants to surround each of these red cubes with a single layer of blue 1 cm³ cubes.

How many blue 1 cm³ cubes will he need to cover each red cube?

Look for a pattern in the answers.

Try to find a general statement.

Teacher's Guide

See page 189 of the *Teacher's Guide* for ideas of how to guide practice. Work through each step together as a class to develop children's conceptual understanding.

175 ⭐

Drawing shapes and finding angles

You need:
- compasses
- geostrips
- split pins
- ruler
- thin card
- scissors

Let's learn

I'm going to rub out the construction arcs that I used to draw my triangle. It will look neater.

Don't do that! They show how you constructed the triangle. It's important to be able to see where the arcs intersect.

Constructing triangles using compasses

How to construct an equilateral triangle with side 5 cm:

- Draw a line exactly 5 cm long.
- Open the compasses to exactly 5 cm.
- Place the point at the end of the line.
  Draw a small arc where you think the apex of the triangle will be.
- Repeat this at the other end of the line.
- The intersection (where the 2 arcs of the circle cross) is the third point of the triangle.
- Carefully draw the other 2 sides of the triangle.

How could you use this method to draw isosceles and scalene triangles?

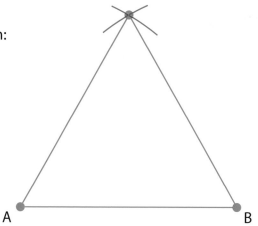

Important angle numbers and facts

90°	Right angle. Internal angle of a rectangle.
180°	Sum of angles on a straight line. Total of angles in a triangle.
360°	Sum of angles around a point. Total of angles in a quadrilateral.
60°	Angle in an equilateral triangle.
45°	Half a right angle.

Teacher's Guide

Before working through the *Textbook*, study page 190 of the *Teacher's Guide* to see how the concepts should be introduced. Read and discuss the page with the children. Provide concrete resources to support exploration.

1 Draw.

Use compasses to construct these triangles:

a   equilateral triangle of side 6.5 cm

b   isosceles triangle with a base of 6 cm and 2 sides of 5 cm.

c   Now measure the angles of the isosceles triangle as accurately as possible.

2 Solve.

Use angle facts and geometry symbols to find the missing angles in these tile patterns.

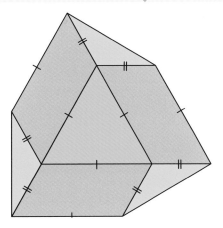
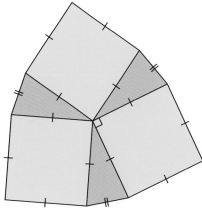

3 Make.

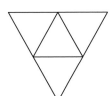
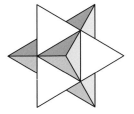

The net of a tetrahedron is 4 equilateral triangles.

Use compasses to construct the net of a tetrahedron of side 10 cm.

Add tabs for sticking. Cut out your net. Make the tetrahedron.

Repeat to make 4 tetrahedrons with sides of 5 cm.

Carefully stick one in the centre of each face of the larger tetrahedron.

4 Investigate.

Find out how to draw a 5-pointed star without removing your pencil from the paper.

This one has a regular pentagon at the centre. There are 5 congruent isosceles triangles.

Using angle facts, calculate the size of each angle.

Calculate the sizes of angles in stars with different numbers of points.

Can you draw any other stars without lifting your pencil?

Teacher's Guide

See page 191 of the *Teacher's Guide* for ideas of how to guide practice. Work through each step together as a class to develop children's conceptual understanding.

177 ★

Reflections and equations

Let's learn

When you say 'reflect a shape in the x-axis', it sounds as if the x coordinates would change sign.

Yes, but think about what happens and look at the coordinate grid. The x coordinate value remains the same and the y coordinate value changes sign.

You need:
- squared paper
- ruler

Reflection in x- and y-axes

Compare the coordinates for triangle A and the coordinates for the reflection in the x-axis, A'.

The general rule for reflection in the x-axis is: $(x, y) \rightarrow (x, -y)$

You can determine the position of the new coordinates for A' using this rule.

The general rule for reflection in the y-axis is: $(x, y) \rightarrow (-x, y)$

You can determine the position of the new coordinates for A" using this rule.

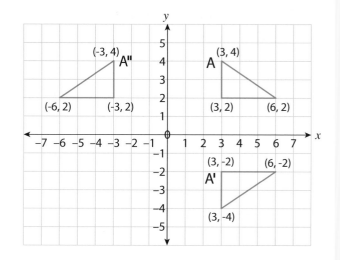

Using an equation to generate coordinates

The equation $y = x + 2$ has two related unknowns, x and y.

When a value for x is chosen, the value for y can be calculated.

When $x = 1$, $y = 3$; when $x = 2$, $y = 4$; when $x = 5$, $y = 7$; when $x = -3$, $y = -1$.

This gives the following coordinate pairs for (x, y): (1, 3), (2, 4), (5, 7), (−3, −1).

When these points are plotted, they lie on a straight line.

The line intersects the y-axis at $y = 2$, when $x = 0$.

It intersects the x-axis at $x = -2$, when $y = 0$.

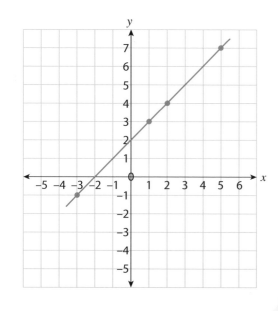

Teacher's Guide

Before working through the *Textbook*, study page 192 of the *Teacher's Guide* to see how the concepts should be introduced. Read and discuss the page with the children. Provide concrete resources to support exploration.

1 Calculate.

Here are the coordinate pairs for a triangle: (2, 2), (6, 7), (4, 1).

Write the new coordinates after:

a   reflection in the x-axis

b   reflection in the y-axis

Check that your points are correct by plotting the original triangle and its reflections.

2 Plot.

Ali plotted where his friends live.
He placed his house at the origin.
Mark and label the location of their homes:

a   Jen (1, −2); Sam (−5, 1); Tom (2, 7); Ana (−3, −2); Lili (4, −9)

b   Reflect the coordinates for Sam's house in the y-axis for Seb's house.

c   Reflect the coordinates for Ana's house in the x-axis for Theo's house.

d   Reflect the coordinates for Ana's house in the y-axis for Amy's house.

e   Translate the coordinates for Jen for Mia's house by −5 (x direction) and −3 (y direction).

f   Translate the coordinates for Tom for Oli's house by −3 (y direction).

g   Translate the coordinates for Sam for Eva's house by +7 (x direction) and −5 (y direction).

h   Who lives closest to Ali and who is furthest away?

3 Apply.

a

x	$y = x - 1$	(x, y)
0	-1 (Think $y = 0 - 1$)	(0, -1)
2		(,)
4		(,)
-1		(,)

b   Plot the coordinate pairs on a full coordinate, join the points and extend the line in both directions. This is the graph of the line $y = x - 1$

c   Use the graph to find the value of x
• when $y = 5$ • when $y = -4$

4 Think.

Here is a simple equation, $y = 2x$.

When $x = 1$, $y = 2$. Draw a table and work out 4 coordinate pairs for the equation.

Plot the coordinate points on a full coordinate grid. Join the points on a line.

Repeat for the equations, $y = x$ and $y = 4x$.

Describe how the lines relate to one another.

Predict where the line $y = 3x$ will fit.

Describe the pattern of the lines.

Teacher's Guide

See page 193 of the *Teacher's Guide* for ideas of how to guide practice.
Work through each step together as a class to develop children's
conceptual understanding.

All about nets

Let's play

Teacher's Guide

See pages 194–5 of the *Teacher's Guide*. Explain the rules for each game and allow children to choose which to play. Encourage them to challenge themselves and practise what they have learnt in the unit.

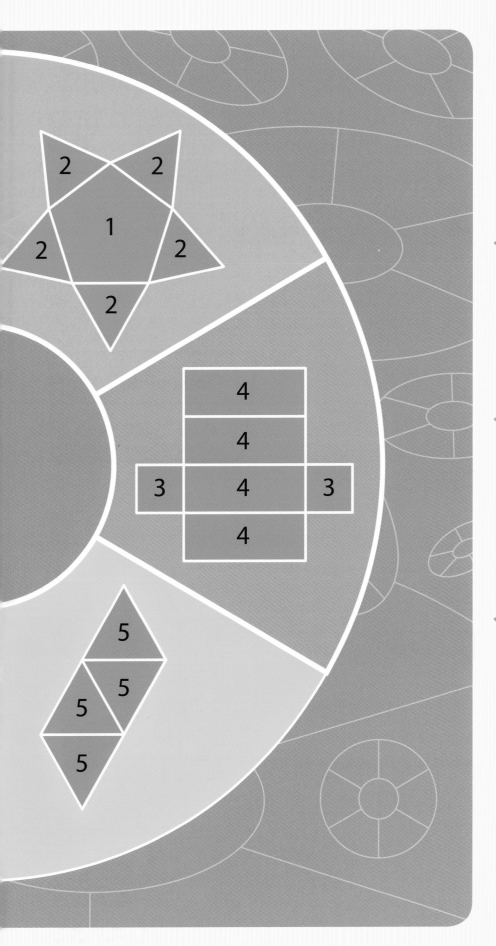

- 1–6 dice
- 16 small counters
- timer

1 Making nets

Choose to make either the orange shapes or the green ones. Take turns to roll the dice. Who can complete their 3 nets first?

2 Angles in nets

Set the timer and take turns to roll the dice. Calculate the score for the angles.

Who can reach the highest score when the time is up?

3 Your game

Make up your own game using the gameboard. Explain the rules and play with a partner.

And finally ...

1

Eva and Ali are looking at some nets of 3-D shapes.
All the nets on Eva's page have triangles in them,
and all the nets on Ali's page have rectangles in them.
Sketch the nets that could be on Eva's and Ali's pages.
What shape appears on both pages?

You need:
- squared paper
- 3-D shapes
- ruler

2

Draw a full coordinate
grid from -10 to +10.

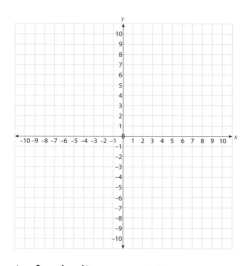

You need:
- squared paper
- ruler

a   Find 3 coordinate pairs for the line $y = x + 2$.
Plot the points and draw the line.
Write down the coordinates of the point where the line crosses the y-axis.

b   Find 3 coordinate pairs for the line $y = x + 5$.
Predict the coordinates of the point where the line crosses the y-axis.
Plot the points to test your prediction.

c   Predict and try some more lines, e.g. $y = x - 3$, $y = x + 7$.

Write a statement
to explain your
findings.

What can you say
about the point
where the line
crosses the x-axis?

Teacher's Guide

See pages 196–7 of the *Teacher's Guide* for guidance on running each task.
Observe children to identify those who have mastered concepts and those who
require further consolidation.

3

Design a symmetrical flag. To do this:
- Draw x-and y-axes from +10 to −10.
- Draw a polygon with 10 or more sides in the first quadrant with each vertex on a point.
- Reflect the shape in the y-axis.
- Reflect the shape and its image in the x-axis.

Choose the same point in each quadrant.
Write the coordinate pair for that point in each quadrant.

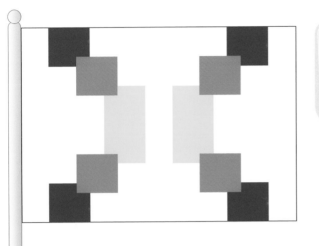

I wonder if there are countries with flags that have horizontal and vertical symmetry like this design?

Did you know?

Look at these amazing hot air balloons! They are made from simple 2-D shapes that are joined to form a net.

They look magical but actually they are a blend of mathematics, science and art. It's a perfect example of beautiful mathematics in action!

⭐6 Glossary

2-dimensional (2-D)

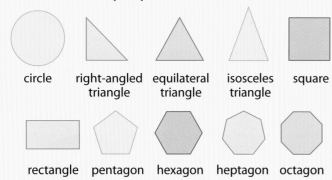

circle   right-angled triangle   equilateral triangle   isosceles triangle   square

rectangle   pentagon   hexagon   heptagon   octagon

3-dimensional (3-D)

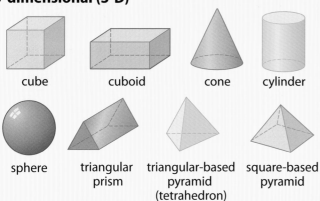

cube   cuboid   cone   cylinder

sphere   triangular prism   triangular-based pyramid (tetrahedron)   square-based pyramid

A

addend

The numbers being added together in an addition calculation. Augend + addend = sum (or total).

$$3 + 5 = 8$$

augend   addend   sum/total

algebra

Where letters or symbols are used for unknown values.

arc

Part of the circumference of a circle.

array

An arrangement of numbers, shapes or objects in rows of equal size and columns of equal size, used to find out how many altogether.

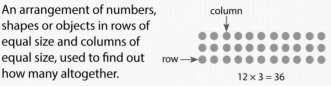

column

row →

$12 \times 3 = 36$

associative

Grouping numbers in different ways to add and multiply:
$5 + 19 + 36 = (36 + 5) + 19 = 41 + 19 = 60$
$4 \times 8 \times 5 = (4 \times 8) \times 5 = 32 \times 5 = 160$

augend

The number being added to in an addition calculation. Augend + addend = sum (or total).

$$3 + 5 = 8$$

augend   addend   sum/total

average

The middle value of a set of numbers. It is found by adding all the numbers together and dividing by how many numbers there are.

B

balance

Things are balanced when both sides have equal value, e.g. $2a + b = c$.

C

capacity

The amount a container holds. It is measured in litres or millilitres, e.g. the capacity of a 2 litre bottle is 2 litres.

centilitre

One hundredth of a litre. Symbol: cl. $100\,cl = 1\,l$.

circumference

The perimeter of a circle. See also *arc*.

commutative

Addition and multiplication are commutative. It doesn't matter which order you add, multiply or divide, the answer is always the same. Same answer, different calculation, e.g. $3 + 4 = 4 + 3$. But subtraction is not commutative, e.g. $7 - 2 \neq 2 - 7$.

concentric

Circles which share the same centre.

congruent

Shapes are congruent if they are exactly the same shape and size.

consecutive

Numbers which follow each other in order.

13, 14, 15
consecutive numbers

24, 26, 28
consecutive even numbers

coordinate

An ordered pair of (x, y) values that gives the position of a point on a graph. In 3-D (x, y, z).

cubic millimetres (mm³), cubic centimetres (cm³), cubic metres (m³), cubic kilometres (km³)

Metric measurements of volume. 1 cm³ is the volume enclosed in a cube of length 1 cm.

cube numbers

Formed when a number is multiplied by itself and then by itself again, e.g. 2 cubed = $2 \times 2 \times 2 = 2^3 = 8$.

D

denominator

The number underneath the vinculum. Also called the divisor.

diameter

A line passing across a circle, or a sphere, which passes through the centre. See also *radius*.

difference

The result of a subtraction. The difference between 12 and 5 is 7. See also *minuend, subtrahend*.

digit total/sum

The sum of all the digits in a number, e.g. the digit sum of 435 is $4 + 3 + 5 = 12$, and $1 + 2 = 3$.

distribution

In statistics. The distribution of a set of values.

distributive law

Multiplying numbers by making equivalent numbers:
$7 \times 12 = (7 \times 7) + (5 \times 7) = 49 + 35 = 84$.
It works for larger numbers too:
$45 \times 6 = (40 \times 6) + (5 \times 6) = 240 + 30 = 270$.

dividend

The number that is divided in a division sum, e.g. in $12 \div 6 = 2$, 12 is the dividend. See also *divisor, quotient*.

$$\overset{\text{dividend}}{\underset{\text{divisor}}{12 \; \div \; 6}} \; = \; 2 \longleftarrow \text{quotient}$$

divisibility

Whether a number can be divided without remainder. All even numbers are divisible by 2.

division bracket

The half box around the dividend in a division. See also *dividend*.

$$16 \overline{)2112} \longleftarrow \text{dividend}$$
division bracket

divisor

The number that is used to divide in a division sum, e.g. in $12 \div 6 = 2$, 6 is the divisor. See also *dividend, quotient*.

dodecahedron

A 3-D polyhedron with 12 faces. A regular dodecahedron has pentagonal faces.

E

equation

A mathematical statement showing an equality, e.g. $10 \times 2 = 4 \times 5$ or $2x + 6 = 16$.

equilateral triangle

A triangle with 3 equal sides and 3 equal angles of 60°.

F

factor

Numbers that divide exactly into a number are its factors, e.g. the factors of 12 are 1, 2, 3, 4, 6, 12.

factorise

To write a number or algebraic expression as a product of 2 or more factors.

foot, feet

An imperial unit of length, approximately 30 cm. 12 inches = 1 foot and 3 feet = 1 yard.

formula, formulae

A mathematical statement using letters or symbols (variables), e.g. Area of a rectangle = length × width or $A = l \times w$.

G

greater than or equal to

Symbol: ≥. An inequality showing the lowest value a number can take. $n \geq 7$ means n can have any value from 7 upwards. See also *less than or equal to*.

I

imperial unit

A unit of measure from pre-metric measurements, e.g. inches, yards, miles, pints. Many are still in common use.

inch, inches

An imperial unit of length, approximately 2.5 cm. 12 inches = 1 foot.

intersecting, intersection

Where two lines cross.

inverse

Inverse operations leave the original value unchanged. The inverse of +4 is – 4. The inverse of × 4 is ÷ 4 or $\times \frac{1}{4}$. The inverse 'undoes' the action.

isosceles triangle

A triangle with 2 equal sides and 2 equal base angles. One of its angles can be a right angle. This is called a right-angled isosceles triangle.

K

kite

A quadrilateral with 2 pairs of equal adjacent sides.

L

less than or equal to

Symbol: ≤. An inequality showing the highest value a number can take. $n \leq 7$ means n can have any value up to and including 7. See also *greater than or equal to*.

linear number sequence

A sequence of numbers that increases by the same difference, e.g. 9, 13, 17, 21, 25 and so on.

M

mean

A measure of average.
Mean = total of all data values ÷ number of data points.

metric unit

Any unit used to measure on a metric scale, e.g. kilograms (kg), centimetres (cm), litres (l). All based on the decimal system.

minuend

The starting number in a subtraction calculation, e.g. 10 (the minuend) – 3 (the subtrahend) = 7 (the difference). See also *subtrahend* and *difference*.

$$10 - 3 = 7 \leftarrow \text{difference}$$
minuend   subtrahend

mixed number

A number with both a whole number part and a fractional part, e.g. $3\frac{1}{2}$.

multiple

A multiple is the product of 2 numbers, e.g. the multiples of 7 are 7, 14, 21, 28 and so on.

multiplicand

A number to be multiplied, e.g. in $6 \times 3 = 18$, 6 is the multiplicand. See also *multiplier*.

$$6 \times 3 = 18$$
multiplicand   multiplier

multiplier

The multiplying number, e.g. in $6 \times 3 = 18$, 3 is the multiplier. See also *multiplicand*.

$$6 \times 3 = 18$$
multiplicand   multiplier

N

net (open, closed)

A pattern that you can cut out and fold to make a 3-D shape.

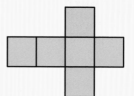

numerator

The number above the vinculum in a fraction. See also *denominator*.

nth term

An unknown value.

O

ounce

An imperial measure of mass. Symbol: oz.
1 ounce is approximately equal to 28 g. 16 oz = 1 pound.

P

parallelogram

A 2-D shape with 2 pairs of opposite sides that are equal and parallel. A rectangle is a parallelogram with all the angles 90°.

pie chart

A circular chart divided into parts.

plane

A flat surface in 2-D.

pound

An imperial measure of mass. Symbol: lb. 16 oz = 1 pound. 2.2 lb is approximately equal to 1 kg. See also *ounce*.

prime factor

A factor of a number that is also a prime number, e.g. the prime factors of 12 are 2 and 3, since $12 = 2 \times 2 \times 3 = 2^2 \times 3$.

product

The result of multiplying 2 numbers.
The product of 4 and 3 is $4 \times 3 = 12$.

profit, loss

The money made or lost in a financial transaction.

Q

quadrant

One of the 4 quarters formed by the x- and y-axes on a graph.

quotient

The answer to a division calculation,
e.g. in $12 \div 6 = 2$, 2 is the quotient. See also *dividend*.

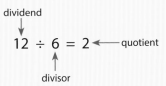

R

radius

Any straight line segment from the centre of a circle to the edge (circumference). The radius is half of the diameter. See also *diameter*.

ratio

A comparison of values or amounts. There are 12 boys for every 15 girls. The ratio is 12 to 15 or 12:15.

reflex angle

An angle greater than 180°.

rhombus

A 2-D shape with 4 equal sides, no right-angles and equal opposite angles.

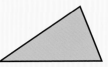

S

scalene triangle

A triangle with no equal sides or angles. A scalene triangle can have a right angle. This is called a right-angled scalene triangle.

statistics

Collecting, representing and interpreting data.

subtrahend

The number that is subtracted from the minuend.

sum

The answer to an addition calculation.
The sum of 4 and 5 is 9. See also *total*.

T

tonne

A metric measure of mass. 1000 kilograms = 1 tonne.

total

The answer to an addition calculation.
The total of 4, 3 and 5 is 12. See also *sum*.

U

unknowns

A symbol for an unknown number, usually a letter.

V

variable

A quantity that we do not know. It can change or may take on different values. A variable is often shown by a letter or symbol, e.g. $3y + 4 = 16$.

vinculum

The line that separates the numerator and denominator in a fraction.

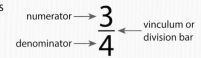

volume

The amount of liquid in a container, e.g. 1 litre of water in a 2 l bottle. Measured in millilitres and litres. See also *capacity*.

W

whole-part relationship

Parts of the whole. In the fraction $\frac{2}{3}$, the whole has been divided into 3 equal parts and we are thinking about 2 of those parts. When thinking of an addition calculation, e.g. $54 + 46 = 100$, 54 and 46 are the parts and 100 is the whole. There are many whole-part relationships in mathematics.

Y

yard

An imperial unit of length. 1 yard is approximately equal to 90 cm. Symbol: yd. 36 inches = 3 feet = 1 yard. See also *foot, feet* and *inch, inches*.